ORIGAMICS
Mathematical Explorations through Paper Folding

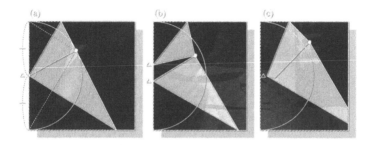

ORIGAMICS
Mathematical Explorations through Paper Folding

Kazuo Haga
University of Tsukuba, Japan

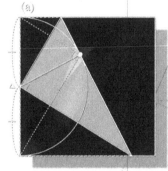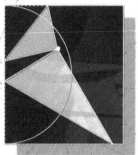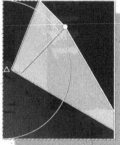

edited and translated by

Josefina C Fonacier
University of Philippines, Philippines

Masami Isoda
University of Tsukuba, Japan

NEW JERSEY · LONDON · SINGAPORE · BEIJING · SHANGHAI · HONG KONG · TAIPEI · CHENNAI

Published by
World Scientific Publishing Co. Pte. Ltd.
5 Toh Tuck Link, Singapore 596224
USA office: 27 Warren Street, Suite 401-402, Hackensack, NJ 07601
UK office: 57 Shelton Street, Covent Garden, London WC2H 9HE

British Library Cataloguing-in-Publication Data
A catalogue record for this book is available from the British Library.

First published 2008
Reprinted 2012

ORIGAMICS
Mathematical Explorations Through Paper Folding

Copyright © 2008 by World Scientific Publishing Co. Pte. Ltd.

All rights reserved. This book, or parts thereof, may not be reproduced in any form or by any means, electronic or mechanical, including photocopying, recording or any information storage and retrieval system now known or to be invented, without written permission from the Publisher.

For photocopying of material in this volume, please pay a copying fee through the Copyright Clearance Center, Inc., 222 Rosewood Drive, Danvers, MA 01923, USA. In this case permission to photocopy is not required from the publisher.

ISBN-13 978-981-283-489-8
ISBN-10 981-283-489-3
ISBN-13 978-981-283-490-4 (pbk)
ISBN-10 981-283-490-7 (pbk)

Printed in Singapore by B & Jo Enterprise Pte Ltd

Introduction

The art of origami, or paper folding, is a great tradition in Japan. In its simplest form, the folding is carried out on a square piece of paper to obtain attractive figures of animals, flowers or other familiar figures. The art enjoys great popularity and appeal among both young and old, and it has spread to other countries beyond Japan.

It is easy to see that origami has links with geometry. Creases and edges represent lines, intersecting creases and edges make angles, the intersections themselves represent points. Because of its manipulative and experiential nature, origami could become an effective context for the learning and teaching of geometry.

In this book, origami is used to reinforce the study of geometry, with the hope that the popularity and appeal for the former will stimulate the latter. The activities in this book differ from ordinary origami in that no figures of objects result. Rather, they lead the reader to study the effects of the folding and seek patterns.

The author, Dr. Kazuo Haga, is a retired professor of biology at the University of Tsukuba, Japan. His interest in science has been channeled to the broader field of science education. He mentioned in his book that during his career as a biology professor, while waiting for his experiments to progress, he used to while away the time doing paper folding (or more specifically, mathematics through paper folding).

The experimental approach that characterizes much of science activity (and possibly much of Professor Haga's work as a biologist) can be recognized throughout the book. The manipulative nature of origami allows much experimenting, comparing, visualizing, discovering and conjecturing. In every topic, the exuberance that the author felt whenever he arrived at mathematical ideas is reflected in his writing style. To paraphrase the author, "more wonders emerge!"

Admittedly proof is a necessary part of mathematical discourse. However, proofs are not emphasized in this book. The author is aware that many students do not appreciate formal proofs. So while some proofs are given after the paper folding, not all mathematical discoveries are proven. The reader is encouraged to fill in all the proofs, for his/her own satisfaction and for the sake of mathematical completeness.

This then is a resource book for mathematics teachers and mathematics teacher educators. It is hoped that going through this book will give them alternative approaches for reinforcing and applying the theorems of high school geometry and for provoking more enthusiasm for mathematics study.

Josefina.C. Fonacier
Former Director,
National Institute for Science and Mathematics Education Development,
University of Philippines

Until the Publication of the English Edition

When I was an undergraduate student almost 30 years ago, our student's mathematics research club, which aimed for understanding mathematics through different ways, held a mathematics exhibition. One of the exhibits was on Origami, paper folding: the mathematics in Orizuru (crane construction), based on the work of Professor Koji Fushimi, a physicist, then the President of the Science Council of Japan. We well remember that Professor Kazuo Haga visited our exhibition, and he explained to us the Haga theorems. He was a biologist, and we were surprised that these works on mathematics and origami had been done by scientists (Fushimi a physicist, and Haga a biologist) and not by mathematicians. Now we are working as mathematics educators in universities, middle schools and high schools, and it was a part of important experience for us in becoming teachers.

When I came back to the University of Tsukuba 15 years ago, Professor Haga began to teach school teachers his mathematical theory of Origami under the name of "ORIGAMICS". I recommended the publisher of the Teachers' Journal on Mathematics Education at Meiji Tosho-Shuppan to have the serial of Professor Haga's "Origamics", because we knew the importance of his activity for mathematics education and teacher education. Based on the series, he published his first book, which would become the major resource of this English translation. Since then he has published two more books. This English translation includes only one third of his works on Origamics.

There are several unique points in his Origamics. The first one comes from the object itself. Everyone has experience in folding a paper, but he explored it based on his unique geometrical ideas. Another point is his approaches in mathematics. He used school mathematics that could be understood by anyone who has studied mathematics at school. Through his mathematical viewpoint, we can learn how to explore and enjoy daily situations geometrically, and develop our mathematical views and minds in the world.

Today, through international conferences, origamics has become a well-known research field throughout the world. Some of Professor Haga's works are lectured by himself at these conferences; at the same time many of his works have been spread through teachers. In the case of the Philippines, Mr. Mikio Masuda, who had been a teacher at the Junior high school (middle school) attached to the University of Tsukuba, was dispatched to the University of the Philippines National Institute for Science and Mathematics Education Development (UP-NISMED) as a specialist of the Japan International Cooperation Agency (JICA) on the appointment of Professor Shizumi Shimizu, University of Tsukuba. Among the people he worked with was Professor Josefina C. Fonacier; she was especially impressed with Professor Haga's work. The major part of this English edition of his book originates from results of their collaborations.

Based on the experience of international cooperation with UP-NISMED through JICA, as well as other international cooperation projects/experiences, the University of Tsukuba established the Center for Research on International Cooperation in Educational Development (CRICED) on behalf of the Ministry of Education, Japan. For developing materials for international cooperation, CRICED staff members have begun fully support for publication. It is my pleasure to edit the English edition of Professor Haga's book with Professor Fonacier, on behalf of my long exchange with Professor Haga and the collaboration experience with UP-NISMED.

Masami Isoda
Center for Research on International Cooperation in Educational Development (CRICED)
University of Tsukuba, JAPAN

Acknowledgments

We would like to acknowledge the following contributors and Institutions:

Dr. Soledad Ulep, Deputy Director of UP-NISMED (the University of the Philippines National Institute for Science and Mathematics Education Development) for her support of our editorial works;

Dr. Yasuo Yuzawa, researcher of CRICED (Center for Research on International Cooperation in Educational Development, University of Tsukuba) for his major contribution for developing editorial version from his mathematical expertise including pictures using Software, Cabri Geometry II+ and LaTeX ;

Dr. Rene Felix, professor at the mathematics department of the University of the Philippines, for his mathematical expertise and care in reading and reviewing the whole manuscript;

Mr. Mikio Masuda, retired teacher of the Junior High School attached to the University of Tsukuba, for his earlier assistance of translation between Professors Haga and Fonacier;

Professor Shizumi Shimizu, the graduate school of human comprehensive science, the University of Tsukuba, for his support to develop international relationship between University of the Philippines and the University of Tsukuba;

Ms. Foo Chuan Eng, Education Officer, Brunei Darussalam for her support of developing captions;

Dr. Hiroshi Yokota, researcher of CRICED for his advice of LaTeX

Programing;

JICA (Japan International Cooperation Agency) and its staff members for providing the supported to establish and develop UP-NISMED and the opportunity to meet and share ideas with professionals from different countries; and UP-NISMED and CRICED staff members who have given full support.

The grant for developing materials under the collaboration between CRICED and JICA from the Ministry of Education, Japan was used for finalizing the part of edited version.

Preface for the English Edition

In 2001, I was given the opportunity to talk about ORIGAMI and ORIGAMICS in the plenary lecture at the Third International Meeting of Origami Science, Mathematics and Education in Asilomar Conference Center, California.

"ORIGAMI" has become an international word at present, such as in Origami Science, which is originally derived from the Japanese word "origami". They differ somewhat in meaning as well as pronunciation. The accent of the former falls on the third syllable (ga) while the latter on the second one (ri), that is, ori[ga]mi and o[ri]gami. Most Japanese pronounce it with a nasal sound. In Japan, origami is usually a handicraft hobby designed mainly for children. Thus almost all the origami books are in the juvenile sections of bookstores, even though some are for enthusiasts and origami scientists. I felt the necessity to give a new name for describing the genre of scientific origami, hence I proposed the term "ORIGAMICS" at the Second Origami Science Meeting in 1994.

The term origamics is composed of the stem "origami" and the suffix "ics", which is often used to indicate science or technology, as in mathematics. Another definite difference between origami and origamics is their end product. The former produces paper models of animals, flowers, fruits, vehicles and so forth; while the latter often does not create beautiful or skillful products, but rather some paper with a lot of wrinkles, furrows or creases.

My first mathematical findings on origami were done in 1978. At that time I was a biologist majoring in arthropodan morphology, and observing tiny insect embryos under a microscope, however, there was no relation between origami mathematics and insect egg study. As the microscopic study needed much time with mental fatigue and eyestrain, I often had a recess and folded a piece of paper torn off from a small notebook for refreshment. Then I discovered some interesting phenomena in the folded paper, and corresponded about them with Professor Koji Fushimi, who was a famous theoretical physicist and also known as origami geometrician. He introduced my findings in the monthly magazine 'Sugaku (Mathematics) Seminar' 18(1):40-41, 1979, titled "Origami Geometry, Haga Theorem" (in Japanese). The theorem was named by him using my surname.

In the subsequent years I discovered several more phenomena on the square and rectangular sheets of paper one after another. The detailed explanations were published in Japanese magazines such as 'Sugaku Seminar' (Nihon-Hyoron-Sha), 'Sugaku Kyoiku' (Meiji Tosho-Shuppan), 'ORU' (Soju-Sha) and 'Origami Tanteidan' (Nihon Origami Gakkai).

I published three books on Origamics, namely: 'Origamics niyoru Sugaku Jugyo' (Meiji Tosho-Shuppan), 1996, 'Origamics Part 1. Geometrical Origami' (Nihonhyoron-sha), 1998 and 'Origamics Part 2. Fold Paper and Do Math' (ditto), 2005. My colleagues recommended me to write an English version of these books. Prof. Josefina C. Fonacier of University of Philippines and Mr. Mikio Masuda of University of Tsukuba (see "Until the Publication of the English Edition") showed special interest and they eagerly drove to translate one of those books into English and to publish it. I responded them and started translation. However, due to my retirement and changing circumstances, I did not managed to complete it. I greatly appreciated and in debt to their kindness and encouragements.

Fortunately, after 10 years of interruption, Associate Professor Masami Isoda of the University of Tsukuba proposed to make a newly edited English version of Origamics as a part of his CRICED activities. I naturally agreed and added new chapters to the original plan. I gave my hearty thanks to Professor Isoda for his proposal and collaboration with Professor Fonanier.

Kazuo Haga

Contents

Introduction   v

Until the Publication of the English Edition   vii

Acknowledgments   ix

Preface for the English Edition   xi

1. A POINT OPENS THE DOOR TO ORIGAMICS   1
   - 1.1 Simple Questions About Origami 1
   - 1.2 Constructing a Pythagorean Triangle 2
   - 1.3 Dividing a Line Segment into Three Equal Parts Using no Tools 5
   - 1.4 Extending Toward a Generalization 8

2. NEW FOLDS BRING OUT NEW THEOREMS   11
   - 2.1 Trisecting a Line Segment Using Haga's Second Theorem Fold 11
   - 2.2 The Position of Point F is Interesting 14
   - 2.3 Some Findings Related to Haga's Third Theorem Fold . 17

3. **EXTENSION OF THE HAGA'S THEOREMS TO SILVER RATIO RECTANGLES**   21

     3.1   Mathematical Adventure by Folding a Copy Paper . . .   21

     3.2   Mysteries Revealed from Horizontal Folding of Copy Paper   25

     3.3   Using Standard Copy Paper with Haga's Third Theorem   30

4. **X-LINES WITH LOTS OF SURPRISES**   33

     4.1   We Begin with an Arbitrary Point   33

     4.2   Revelations Concerning the Points of Intersection . . .   35

     4.3   The Center of the Circumcircle!   37

     4.4   How Does the Vertical Position of the Point of Intersection Vary?   38

     4.5   Wonders Still Continue   41

     4.6   Solving the Riddle of "$\frac{1}{2}$"   42

     4.7   Another Wonder   43

5. **"INTRASQUARES" AND "EXTRASQUARES"**   45

     5.1   Do Not Fold Exactly into Halves   46

     5.2   What Kind of Polygons Can You Get?   46

     5.3   How do You Get a Triangle or a Quadrilateral?   48

     5.4   Now to Making a Map   49

     5.5   This is the "Scientific Method"   53

     5.6   Completing the Map   53

     5.7   We Must Also Make the Map of the Outer Subdivision   55

     5.8   Let Us Calculate Areas   57

6. A PETAL PATTERN FROM HEXAGONS? 59

6.1 The Origamics Logo 59

6.2 Folding a Piece of Paper by Concentrating the Four Vertices at One Point 60

6.3 Remarks on Polygonal Figures of Type n 63

6.4 An Approach to the Problem Using Group Study 64

6.5 Reducing the Work of Paper Folding; One Eighth of the Square Will Do 65

6.6 Why Does the Petal Pattern Appear? 66

6.7 What Are the Areas of the Regions? 70

7. HEPTAGON REGIONS EXIST? 71

7.1 Review of the Folding Procedure 71

7.2 A Heptagon Appears! 73

7.3 Experimenting with Rectangles with Different Ratios of Sides 74

7.4 Try a Rhombus 76

8. A WONDER OF ELEVEN STARS 77

8.1 Experimenting with Paper Folding 77

8.2 Discovering 80

8.3 Proof 82

8.4 More Revelations Regarding the Intersections of the Extensions of the Creases 85

8.5 Proof of the Observation on the Intersection Points of Extended Edge-to-Line Creases 89

8.6 The Joy of Discovering and the Excitement of Further Searching 91

9. WHERE TO GO AND WHOM TO MEET 93

9.1 An Origamics Activity as a Game 93

9.2 A Scenario: A Princess and Three Knights? 93

9.3 The Rule: One Guest at a Time 94

9.4 Cases Where no Interview is Possible 97

9.5 Mapping the Neighborhood 97

9.6 A Flower Pattern or an Insect Pattern 99

9.7 A Different Rule: Group Meetings 99

9.8 Are There Areas Where a Particular Male can have Exclusive Meetings with the Female? 101

9.9 More Meetings through a "Hidden Door" 103

10. INSPIRARATION OF RECTANGULAR PAPER 107

10.1 A Scenario: The Stern King of Origami Land 107

10.2 Begin with a Simpler Problem: How to Divide the Rectangle Horizontally and Vertically into 3 Equal Parts 108

10.3 A 5-parts Division Point; the Pendulum Idea Helps . . 111

10.4 A Method for Finding a 7-parts Division Point 115

10.5 The Investigation Continues: Try the Pendulum Idea on the 7-parts Division Method 117

10.6 The Search for 11-parts and 13-parts Division Points 120

10.7 Another Method for Finding 11-parts and 13-parts Division Points 122

10.8 Continue the Trend of Thought: 15-parts and 17-parts Division Points 125

10.9 Some Ideas related to the Ratios for Equal-parts Division based on Similar Triangles 130

10.10 Towards More Division Parts 134
10.11 Generalizing to all Rectangles 134

Topic 1

A POINT OPENS THE DOOR TO ORIGAMICS

Haga's First Theorem and its Extensions

1.1 Simple Questions About Origami

Whenever an origami activity is brought up in the classroom the students show great interest and enthusiasm. And even as the colored pieces of origami paper are distributed, the students are in a hurry to start some folding process. This burst of eagerness of the students allows for a smooth introduction to the subject matter of this topic.

As the students make their first fold, call their attention to all the first folds. The objects the students plan to make may vary - flower, animal or whatever. But no matter what they are trying to make, their first fold is invariably one of these types: a book fold (or side-bisector fold) made by placing one side on the opposite side and making a crease as in Fig.1.1(a), or a diagonal fold made by placing a vertex on the opposite vertex and making a crease as in Fig.1.1(b). In the book fold two opposite sides are bisected, hence the alternate name.

Why are there only two types? We explain this. In origami all acceptable folds must have the property of reproducibility - the result of a folding procedure must always be the same. The basic origami folds involve point-to-point or line-to-line. Using only the four edges and the four vertices, the possible ways of folding are placing an edge onto another edge or placing a vertex onto another vertex. By considering all such manipulations one sees that the only possible outcomes are the two folds mentioned above.

2   Haga's First Theorem and its Extensions

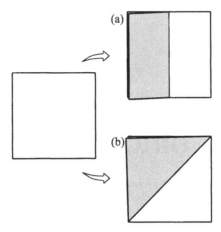

Fig. 1.1   The first folds with the property of reproducibility.

While musing over the above observations one might ask: what other folds are possible if, in addition to the four vertices, another point on the square piece of paper were specified? This question plants the seed for the discussions in this book and opens the door to origamics - classroom mathematics through origami.

1.2 Constructing a Pythagorean Triangle

When we are told to select a particular point on the square paper other than the vertices without using any tool (that is, no ruler or pencil), the simplest to be selected is the midpoint of a side. To mark a midpoint start bending the paper as for a book fold, but do not make a full crease. Just make a short crease on the edge of the square or make a short mark with one's fingernails. It is not necessary to make a crease the whole length of the paper; too many crease marks are likely to be an obstacle to further study. We shall call a small mark like this a scratch mark or simply a mark.

Now we make a fold on the paper with this midpoint as reference or starting point. Several methods of folding can be devised. One folding method is to place a vertex on the mark, another folding method is to make a crease through the mark. The method should be such that a

A POINT OPENS THE DOOR TO ORIGAMICS

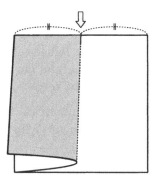

Fig. 1.2 Make a small mark on the midpoint of the upper edge.

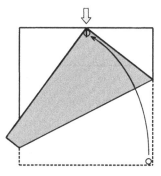

Fig. 1.3 Place the lower right vertex onto the midpoint mark.

unique fold is obtained, no matter how often or by whom it is made.

In this topic we shall discuss one folding procedure and some properties related to it. Other ways of folding shall be discussed in other topics.

To facilitate discussion let us set the standard position of the square piece of paper to be that where the sides are horizontal (that is, left to right) or vertical (that is, upwards or downwards). Therefore we shall designate the edges as left, right, upper, or lower; and the vertices as upper left, lower left, lower right, or upper right.

Select the midpoint of the upper edge as starting point (Fig.1.2). Place the lower right vertex on the starting point and make a firm crease (Fig.1.2). Either the right or left lower vertex may be used, it does not make any difference for analysis purposes. But to follow the diagrams we shall use the lower right vertex.

By this folding process a non-symmetrical flap is made. A number of interesting things can be found about it. To facilitate discussion, in Fig.1.4 points were named.

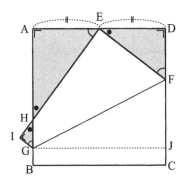

Fig. 1.4   There are three similar right angled triangles.

Let the length of one side of the square be 1.

First, in right $\triangle$DEF we can find the lengths of the sides. Let DF $= a$. Then FC $= 1 - a$. By the folding process FE $=$ FC, so FE $= 1 - a$. Since E is a midpoint, DE $= \dfrac{1}{2}$. Applying the Pythagorean relation, $(1-a)^2 = a^2 + \left(\dfrac{1}{2}\right)^2$. From this we obtain $a = \dfrac{3}{8}$. Therefore DF $= \dfrac{3}{8}$ and FE$= 1 - a = \dfrac{5}{8}$. In other words by the above folding procedure the right side of the square is divided in the ratio 3 : 5. And further, the ratio of the three sides of $\triangle$EDF is

$$FD : DE : EF = \frac{3}{8} : \frac{1}{2} : \frac{5}{8} = 3 : 4 : 5.$$

$\triangle$EDF turns out to be a *Pythagorean Triangle!*

Such triangles were used by the Babylonians, the ancients Egyptians such as for land surveying along the lower Nile River and the ancients Chinese. History tells us that several thousand years ago there was repeated yearly flooding of the river, so land boundaries were continually erased. For resurveying these boundaries they made use of the Pythagorean triangle. The 3 : 4 : 5 triangle is often mentioned as the origin of geometry.

Constructing the Pythagorean triangle by Euclidean methods - that is, with the use of straight edge and compass - requires a lot of time. By contrast, as you have seen, this can be done in origamics with just one fold on the square piece of paper.

1.3 Dividing a Line Segment into Three Equal Parts Using no Tools

Still other triangles emerge from the folding procedure. The lengths of their sides reveal some interesting things.

We determine the lengths of the sides of $\triangle$EAH in Fig.1.4. As before, let the length of the side of the square be 1. Since vertex C of the square was folded onto point E and C is a right angle, then also HEF is a right angle. Therefore the angles adjacent to $\angle$HEF are complementary and $\triangle$EAH and $\triangle$FDE are similar. Therefore $\triangle$AEH is also an Egyptian triangle.

Now we look for AH. By the proportionality of the sides

$$\frac{DF}{DE} = \frac{AE}{AH}, \quad \text{then} \quad \frac{\frac{3}{8}}{\frac{1}{2}} = \frac{\frac{1}{2}}{AH}.$$

Therefore $AH = \frac{2}{3}$.

This value of AH is another useful surprise. It indicates that by locating the point H one can find $\frac{1}{3}$ of the side - BH is $\frac{1}{3}$ of the side. That is, H is a trisection point.

Dividing a strip of paper into three equal parts is often done by lightly bending the strip into three parts and shifting these parts in a trial-and-error fashion until they appear equal. Because trial-and-error is involved this method is imprecise and therefore is not mathematically acceptable. Other trisecting methods by origami have been reported, but the method above described is one of the simplest and neatest. In fact, it is possible to carry out the procedure of marking the trisection point with almost no creases.

We continue to look for the other sides of $\triangle$AEH. We look for side HE.

$$\frac{DF}{EF} = \frac{AE}{HE}, \quad \text{then} \quad \frac{\frac{3}{8}}{\frac{5}{8}} = \frac{\frac{1}{2}}{HE}.$$

Therefore $HE = \frac{5}{6}$.

This value of HE is also useful in that it enables us to find $\frac{1}{6}$ of the side. By returning the flap to the original position EH falls on side CB, so that H separates $\frac{1}{6}$ of the side. That is, H is a *hexasection* point of the side.

There is still another triangle to study in Fig.1.4, right angled $\triangle$GIH. Since $\angle$GHI and $\angle$EHA are vertical angles and are therefore equal, then $\triangle$GIH and $\triangle$EAH are similar. So $\triangle$GIH is still another Egyptian triangle with

$$GI : IH : HG = 3 : 4 : 5.$$

Also, since $EI = CB = 1$, then $HI = EI - EH = 1 - \frac{5}{6} = \frac{1}{6}$. As for the other sides of $\triangle$GIH, it is easy to obtain $GI = \frac{1}{8}$ and $GH = \frac{5}{24}$.

Finally, to complete our study of the segments in Fig.1.4, we look for the length of FG. Imagine a line (or fold) through G parallel to the lower edge BC and intersecting side CD at point J. This line forms a right $\triangle$FJG with hypotenuse FG. Since by folding $GB = GI$, then $GI = JC = \frac{1}{8}$ and $JF = CF - CJ = \frac{5}{8} - \frac{1}{8} = \frac{1}{2}$. Therefore by applying the Pythagorean Theorem to $\triangle$FJG, $FG = \frac{\sqrt{5}}{2}$.

The main ideas just discussed are summarized in the following theorem.

A POINT OPENS THE DOOR TO ORIGAMICS

> **Haga's First Theorem** By the simple folding procedure of placing the lower right vertex of a square onto the midpoint of the upper side, each edge of the square is divided in a fixed ratio, as follows (see Fig.1.5).
>
> (a) The right edge is divided by the point F in the ratio 3 : 5.
> (b) The left edge is divided by the point H in the ratio 2 : 1.
> (c) The left edge is divided by the point G in the ratio 7 : 1.
> (d) The lower edge is divided by the point H in the ratio 1 : 5.

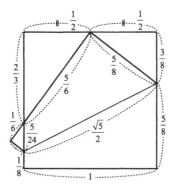

Fig. 1.5   Various lengths appears by folding only once.

And the fold used in the theorem is called **Haga's First Theorem Fold**.

In (a) and (c) the ratios may be obtained by dividing a side in half, then again in half, then still again in half (that is, dividing the side into 8 equal parts). But the ratios in (b) and (d) cannot be so obtained. For this reason this one-time folding method is a useful, simple and precise dividing procedure.

Comment. The discoveries described in this topic were first reported as "Haga's Theorem" by Dr. Koji Fushimi in the journal Mathematics Seminar volume 18 number 1 (January 1979, in Japanese). Other folding methods have since been explored by Haga, hence the change in name in 1984 to "Haga's First Theorem".

8   Haga's First Theorem and its Extensions

Dr. Fushimi is a past chairman of the Science Council of Japan. He is author of "Geometrics of Origami" (in Japanese) published by the Nippon Hyoronsha.

1.4 Extending Toward a Generalization

So far the folding procedures have been based on the midpoint of an edge as starting point. We might ask ourselves: what results would we obtain if the starting point were some other point on the edge?

In Fig.1.6 an arbitrary point was chosen and is indicated by an arrow. In Fig.1.7 the vertices of the squares are named. Denote the chosen point by E and the distance DE by x. Denote the different segments by y_1 to y_6 as in Fig.1.7. Then the lengths of the segments become functions of x as follows.

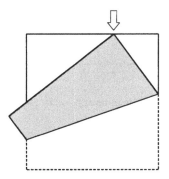

Fig. 1.6 Folding onto positions other than the midpoint.

[y_1]   By the Pythagorean relation on $\triangle$DEF, $x^2 + y_1^2 = (1 - y_1)^2$. So
$$y_1 = \frac{1 - x^2}{2} = \frac{(1+x)(1-x)}{2}.$$

[y_2]   Since $\triangle$AHE is similar to $\triangle$DEF, then $\dfrac{y_1}{1-x} = \dfrac{x}{y_2}$. So
$$y_2 = \frac{2x}{1+x}.$$

[y_3]   Also from similar triangles $\triangle$AHE and $\triangle$DEF, we obtain
$$\frac{y_2}{y_3} = \frac{x}{1 - y_1}. \text{ So } y_3 = \frac{1 + x^2}{1 + x}.$$

A POINT OPENS THE DOOR TO ORIGAMICS

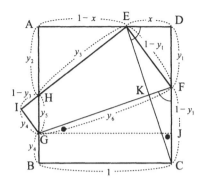

Fig. 1.7 y_1 to y_6 indicates the other various lengths.

[y_4] Since FG and EC are perpendicular, $\triangle$CKF and $\triangle$CDE are similar. Therefore $\angle$DEC and $\angle$KFC are congruent; so also $\triangle$CDE and $\triangle$GJF. So FJ = x. Therefore $y_4 = \text{JC} = 1 - (y_1 + x) = \dfrac{(1-x)^2}{2}$.

[y_5] Since $y_2 + y_5 + y_4 = 1$, $y_5 = 1 - \left(\dfrac{2x}{1+x} + \dfrac{(1-x)^2}{2}\right)$.

[y_6] By the Pythagorean relation on $\triangle$GJF, $y_6 = \sqrt{\text{FJ}^2 + \text{JG}^2} = \sqrt{x^2 + 1}$.

It is difficult to feel excited over the above relations if described only in terms of formal general expressions. To help us better appreciate these relations let us find their values for particular values of x. Using the square pieces of paper, locate the points corresponding to $x = \dfrac{1}{4}$ and $x = \dfrac{3}{4}$. We fold as before, placing the lower vertex on each mark as in Figs.1.8(a) and (b).

The values of the y's for these two values of x, as well as those for $x = \dfrac{1}{2}$, are given in the table below.

From the table we see that various fractional parts are produced, the simpler ones being halves, thirds, fourths, fifths, sixths, sevenths and eighths. We realize that by selecting suitable values of x we can obtain segments of any fractional length or their integral multiples. Therefore with no tools, by simply marking a specific dividing point on an edge and making just one fold, any fractional part of the square

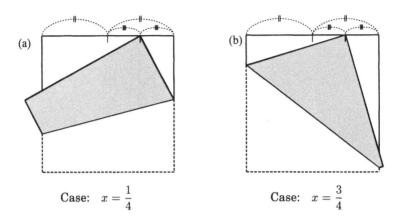

Case: $x = \dfrac{1}{4}$ Case: $x = \dfrac{3}{4}$

Fig.1.8

piece of paper may be obtained. And to reduce the clutter of too many folds, just fold in parts (i.e., small scratch marks instead of whole creases) to obtain the important points.

x	$\dfrac{1}{2}$		$\dfrac{1}{4}$		$\dfrac{3}{4}$	
y_1 / $1-y_1$	$\dfrac{3}{8}$	$\dfrac{5}{8}$	$\dfrac{15}{32}$	$\dfrac{17}{32}$	$\dfrac{7}{32}$	$\dfrac{25}{32}$
y_2 / $1-y_2$	$\dfrac{2}{3}$	$\dfrac{1}{3}$	$\dfrac{2}{5}$	$\dfrac{3}{5}$	$\dfrac{6}{7}$	$\dfrac{1}{7}$
y_3 / $1-y_3$	$\dfrac{5}{6}$	$\dfrac{1}{6}$	$\dfrac{17}{20}$	$\dfrac{3}{20}$	$\dfrac{25}{28}$	$\dfrac{3}{28}$
y_4 / $1-y_4$	$\dfrac{1}{8}$	$\dfrac{7}{8}$	$\dfrac{9}{32}$	$\dfrac{23}{32}$	$\dfrac{1}{32}$	$\dfrac{31}{32}$

Thus, in spite of the austere simplicity of this "one-fold" procedure, many exciting revelations emerge. Clearly Haga's First Theorem Fold is highly worthwhile.

Topic 2

NEW FOLDS BRING OUT NEW THEOREMS

Mathematical Principles Related to Haga's Second and Third Theorems

2.1 Trisecting a Line Segment Using Haga's Second Theorem Fold

In the previous topic, just by placing a lower vertex of a square piece of paper on the midpoint of the upper edge and making a crease many interesting ideas about the resulting segments and angles came to light. One such idea was discussed in topic 1 as "Haga's First Theorem", and the related fold was named "Haga's First Theorem Fold" (Fig.2.1(a)).

In the present topic we shall discuss "Haga's Second Theorem Fold". As before mark the midpoint of the upper edge of the paper. Then make the fold linking this midpoint and a vertex of the bottom edge. This is an unusual way of folding and you may find it a bit difficult; so fold very carefully, especially when folding through the midpoint (Fig.2.1(b)). Make a firm crease.

In Fig.2.2 points were named: E is the midpoint of side AD of the square ABCD. EC is the resulting crease and right $\triangle$EFC is the resulting triangular flap. If the length of a side of the square is 1, then the length of the crease EC is $\frac{\sqrt{5}}{2}$, obtained by using the Pythagorean theorem on right $\triangle$EDC.

We shall study the folded part of the top edge, EF in Fig.2.2. This is the new position of ED after folding. Suppose segment EF is extended to the left edge at point G. Where does this extended line reach on the

12 *Mathematical Principles Related to Haga's Second and Third Theorems*

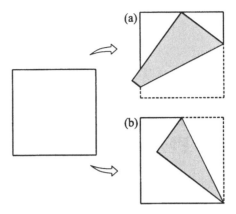

Fig. 2.1   How to fold the upper edge through the midpoint?

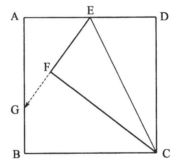

Fig. 2.2   Where is the position of G, when the extended line EF meets line AB?

left side?

When I folded this way the first time, I intuitively sensed that the point G would be a trisection point of the left side (because I was studying this right after I found the trisection point by Haga's First Theorem fold).

Comment on teaching.   On a few occasions I asked several of my students the following question:

When a piece of origami paper is folded as shown in Fig.2.2, and the line EF is extended to meet the left side at point G, it seems that the length of BG is one third of the left side. Is this true?

It took some my college students (biology majors) a long time to answer the question; others gave up. It took some middle school students less time to solve this. One entomologist from a foreign country who was visiting Japan sent me his answer one month after his return to his home country.

To help us answer the question fold the paper once again as in Fig.2.3. That is, place the bottom edge BC on the edge EC of the triangular flap to make a second triangular flap. Since these two edges are sides of the square paper their lengths are the same, and since both adjoining vertices are right angles the two short legs of the triangles are collinear. Therefore the end of the fold is point G.

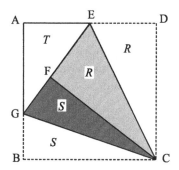

Fig. 2.3 Two pair of similar right angled triangles are used to prove G is a trisection point.

Denote the areas of the triangular flaps as R and S respectively, and the area of $\triangle$AEG as T. Then the area of the whole square is 2R+2S+T. Assuming that the length of one side of the square is 1 and letting BG $= x$, then

$$R = \frac{1}{4}, \quad S = \frac{x}{2}, \quad T = \frac{1-x}{4}.$$

Therefore the expression for the area of the square becomes

$$2\left(\frac{1}{4}\right) + 2\left(\frac{x}{2}\right) + \frac{1-x}{4} = 1.$$

From this equation we obtain $x = \frac{1}{3}$.

So we made a good guess that the point G is a trisection point.

There are other ways to prove that G is a trisection point; think of other proofs. A proof is possible without use of the Pythagorean theorem. Constructing a second proof could be an exercise for second grade middle (8th grade) students.

> **Haga's Second Theorem** Mark the midpoint of the upper edge of a square piece of paper, and make a crease through this midpoint and the lower right vertex. A right triangular flap is formed. If the line of the shorter leg of the flap is extended to intersect the left edge of the square, the intersection point divides the left edge into two parts, the shorter part $= \frac{1}{3}$ of the whole edge (Fig.2.2).

Comment. This theorem was introduced by Mr. Yasuhari Hushimi as "Haga's Second Theorem" in the extra issue Origami no Kagaku (Science of Origami) attached to the Journal of Science, October 1980, published by the Nihon Keizai Shinbun-sya. And the fold produced was named the "Second Theorem Fold".

2.2 The Position of Point F is Interesting

The position of the point F may be arrived at by using ratios of corresponding sides of similar triangles.

See Fig.2.4. Let the intersection of DF and EC be H, and let I be the foot of the perpendicular from F to the upper edge. The right $\triangle$CDE and $\triangle$DHE have a common acute angle, and right $\triangle$DHE and $\triangle$DIF have a common acute angle, therefore the three triangles are similar [1].

From $\triangle$CDE $\sim$ $\triangle$DHE, DH$= \frac{1}{\sqrt{5}}$. So DF$= $ 2DH $= \frac{2}{\sqrt{5}}$.

From $\triangle$CDE $\sim$ $\triangle$DIF, CD : CE $=$ DI : DF.

Then $1 : \frac{\sqrt{5}}{2} =$ DI $: \frac{2}{\sqrt{5}}$. So DI $= \frac{4}{5}$.

---
[1] Here, Similar is represented by the symbol '$\sim$'.

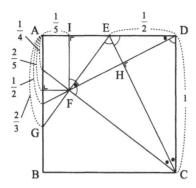

Fig. 2.4 Various lengths could be obtained by the Second Theorem fold. Various lengths could be obtained by the Second Theorem fold.

Therefore $AI = \frac{1}{5}$; that is, point F is $\frac{1}{5}$ from the left edge.

From $\triangle CDE \sim \triangle DIF$, $DI : FI = 2 : 1$. So $FI = \frac{2}{5}$; that is, point F is $= \frac{2}{5}$ from the upper edge.

Comment. This property was discovered by Mr. Kunihiko Kasahara.

Thus, by just marking the midpoint of the upper side of the square paper and making a Second Theorem fold, one can easily obtain $\frac{1}{5}, \frac{2}{5}, \frac{3}{5}$ or $\frac{4}{5}$ of the side of the square. You cannot obtain these results as easily by using compass and straightedge.

Comment on teaching. As a classroom activity, how a teacher develops the topic depends on his/her style. However, before bringing up a mathematical proof it would be more concrete and interesting for the students if a folding procedure to support the above findings is brought up first. Following is a suggestion. The point F is first located by Haga's Second Theorem Fold. The teacher then proposes that the position of F is as stated above. Then he/she asks for paper folding procedures to support the proposition. Finally, after paper folding, he/she develops a proof.

A sample folding procedure to verify that F is $\frac{1}{5}$ from the left side is as follows (Figs.2.5(a) and (b)). On a piece of square paper mark the point F (Haga's second theorem fold).

(1) Place side CD onto F to make a vertical fold. This makes a rectangular flap. Unfold.
(2) Place side CD onto the vertical fold of step (1). This makes a rectangular flap and a second vertical fold. Unfold.
(3) Move side AB onto the vertical fold of step (2). This makes a rectangular flap and a third vertical fold. Unfold.
(4) Divide the rectangular flap of step (3) into two equal parts to make a fourth vertical fold. This last fold should pass through F.

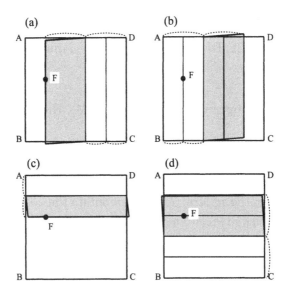

Fig. 2.5 5 equal division of the paper can be produced by using point F.

A sample folding procedure to show that F is $\frac{2}{5}$ from the top edge is shown in Fig.2.5(c) and (d).

The folding procedures may be repeated with 2 or 3 other pieces of paper to be more certain of the results. Finally before the end of the lesson a mathematical proof should be developed.

2.3 Some Findings Related to Haga's Third Theorem Fold

Two folds have been brought up of folding a square piece of paper using the midpoint of the upper edge as reference point. There is still another way of folding.

Comment. I thought of this folding more than 10 years after the publication of Haga's Second Theorem. My profession is biology, and for a while I concentrated on a new phenomenon in my field, and so I did not have time to play with those paper squares. I used to do paper folding while riding on a bus or train. I thought of this new way of folding while I was on a bus going from Tokyo station to Tsukuba Center. I did not notice that I was talking to myself at that time, until some of the other passengers started staring at me. Suddenly I felt embarrassed. Nevertheless, I was excited over my discovery and continued to look for relations as the bus proceeded.

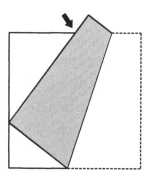

Fig. 2.6   There is yet another way of folding the paper onto the midpoint of the upper edge.

The new folding procedure is shown in Fig.2.6. The starting point is also the midpoint of the upper edge. Mark this midpoint. Then lightly bend the paper so that the right vertex falls on the left edge; do not

18 Mathematical Principles Related to Haga's Second and Third Theorems

make a crease. Shift this lower vertex along the left edge until the right edge is on the marked midpoint. Hold this position and make a firm crease. Remember: the lower vertex should be on the left edge and the right edge should pass through the marked midpoint.

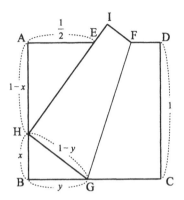

Fig. 2.7   The ideas of the clarification of the Third Theorem fold.

It appears that the intersection on the left edge cuts off $\frac{1}{3}$ of the edge. We proceed to prove that this is indeed true.

With the help of Fig.2.7 we shall prove that HB $= \frac{1}{3}$.

Assume that the length of the side of the square is 1. Let HB $= x$ and BG $= y$. By the paper folding procedure

$$CG = GH = 1 - y.$$

Applying the Pythagorean theorem to $\triangle$HBG,

$$x^2 + y^2 = (1-y)^2,$$

and therefore

$$y = \frac{1-x^2}{2}. \tag{1}$$

Since $\angle$EHG is right, then $\angle$AHE and $\angle$BHG are complementary. Therefore $\triangle$EAH and $\triangle$HBG are similar, and AE : AH = HB : BG, or $\frac{1}{2} : 1 - x = x : y$.
This results in

$$y = 2x(1-x).$$

Substituting in (1) we obtain
$$\frac{1-x^2}{4} = x - x^2,$$
which leads to
$$3x^2 - 4x + 1 = 0 \quad \text{or} \quad (3x-1)(x-1) = 0.$$
The roots are $x = 1$ and $x = \frac{1}{3}$. The value $x = 1$ is discarded, so $x = \frac{1}{3}$.

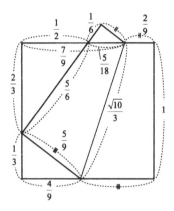

Fig. 2.8 By the Third Theorem fold, not only $\frac{1}{3}$ but also the segments of $\frac{1}{6}$ and $\frac{1}{9}$ are possible.

Comment. It was fortuitous that in 1994 the Second World Conference on the Science of Origami was held. At this conference I gave a talk on the opening day wherein I announced this new theorem. And I named it Haga's Third Theorem, following the First and Second theorems.

Haga's Third Theorem Mark the midpoint of the upper edge of a square piece of paper. Bend the paper to place the lower right vertex on the left edge, then shift it upward or downward until the right edge of the paper passes through the marked midpoint. Make a firm crease. The crease formed divides the left edge into two parts, the shorter part $\frac{1}{3}$ of the whole edge.

In the course of the proof I solved for the lengths of the parts of all the sides. Figure 2.8 shows these lengths. Notice that we get lengths of, not only thirds, but also sixths and ninths. You can see how easily these fractional parts were obtained by paper folding. By contrast, these fractional parts are not easy to arrive at if you use straight edge and compass only.

Topic 3

EXTENSION OF THE HAGA'S THEOREMS TO SILVER RATIO RECTANGLES

Vertical and Horizontal Layouts

3.1 Mathematical Adventure by Folding a Copy Paper

Square paper origami is easily available, easy to become familiar with, and easy to use; and because of this one somehow feels that square origami is play. The author therefore decided to use another kind of paper, also used frequently in everyday life. Copy paper, note paper, writing paper, report paper, memo notepads, and the like may have sides in the ratio $1 : \sqrt{2}$. This ratio is important because when such rectangles are folded in half, the resulting all display the same ratio $1 : \sqrt{2}$.[2]

Size A4 paper is popular in offices. Take a sheet of A4 paper. To verify that the sheet has the required ratio of sides, fold a square on one end of the rectangle with side the short end of the rectangle. Then fold the square to obtain the diagonal (note that the diagonal has length $\sqrt{2}$). The long side of the rectangle should be the same length as the diagonal of the square. There are two ways to position the rectangular sheet of paper. One way is when the long side is in the vertical position as in Fig.3.1; another is such that the long side is horizontal as in Fig.3.4. We

---

[2]Most countries have adopted the international standard for paper sizes, paper with ratio of sides $1 : \sqrt{2}$. An important property of such standard rectangles is that if you divide a rectangle crosswise into two equal parts the resulting rectangles will also have the same ratio of sides. Thus for the sequence of A sizes, A0, A1, A2, A3, A4, ... each size is half the preceding size in area but all have ratios of sides $1 : \sqrt{2}$. Rectangles with this ratio of sides are also called *silver rectangles*.

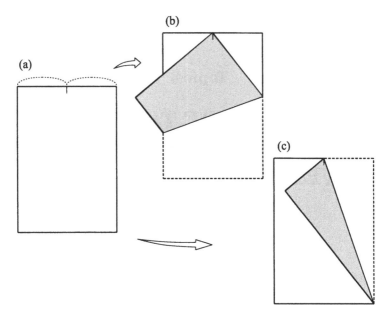

Fig. 3.1
(a) Rectangular paper in a vertical layout,
(b) Extension of Haga's first theorem, and
(c) Extension of Haga's second theorem

shall call these paper positions the vertical and the horizontal layout respectively. First place the rectangle in vertical position as in Fig.3.1. Let us now construct Haga's First Theorem Fold. Mark the midpoint of the top edge and fold the paper so that one vertex of the lower side coincides with this midpoint (the lower right vertex in Fig.3.1(b)).

We can now calculate different lengths. Be aware that the sides of the rectangle have different lengths so that a calculated length may be a fractional part of the short side or a fractional part of the long side. In Fig.3.2 (and the other figures in this topic), fractions referring to a part of the long side are in italics; fractions referring to the short side are in ordinary print. For example, segment HE is $\frac{9}{14}$ of the short side; EF is $\frac{9}{16}$ of the long side or $\frac{9}{16}$ of $\sqrt{2}$.

The calculations are left to the reader; the results are shown in Fig.3.2. From this figure note that

EXTENSION OF THE HAGA'S THEOREMS TO SILVER RATIO RECTANGLES

In the following figure, fractions in italics, $\frac{b}{a}$, represent the ratio when the long side is one unit.

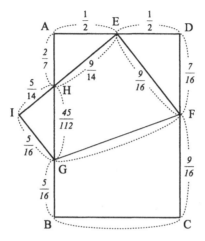

Fig. 3.2 Lengths obtained by Haga's first theorem of rectangle in a vertical layout.

FD is $\frac{7}{16}$ of CD, or CD is divided in the ratio 7 : 9 by G;

AF is $\frac{11}{16}$ of AB, or AB is divided in the ratio 11 : 5 by F;

AH is $\frac{2}{7}$ of AB, or AB is divided in the ratio 2 : 5 by H;

and furthermore, for the bottom side, which was folded back to the position IE,

BC is divided in the ratio 5 : 9 by H.

Thus, points F, G and H can be used to divide the rectangle into 16ths, 9ths and 14ths respectively.

Among these four ratios, of particular importance are the last two. The first two ratios imply division into 16 parts, such division easily obtained by folding into two equal parts, then again into two equal parts, repeatedly folding again into two equal parts to finally obtain 16 equal parts. But the relationships displayed by the last two imply division into 7 or 14 equal parts; and such divisions not obtainable by folding in any simple manner.

The calculations are left to the reader; the results are shown in Fig.3.2. From this figure note that

FD is $\frac{7}{16}$ of CD, or CD is divided in the ratio 7 : 9 by G;

AG is $\frac{11}{16}$ of AB, or AB is divided in the ratio 11 : 5 by F;

AH is $\frac{2}{7}$ of AB, or AB is divided in the ratio 2 : 5 by H;

and furthermore, for the bottom side, which was folded back to the position IE,

BC is divided in the ratio 5 : 9 by H.

Thus, points F, G and H can be used to divide the rectangle into 16ths, 9ths and 14ths respectively.

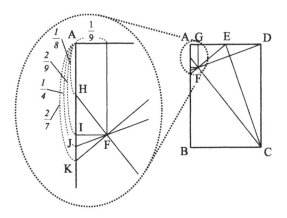

Fig. 3.3 Lengths according to point F of a rectangle in a vertical layout using Haga's second theorem.

Among these four ratios, of particular importance are the last two. The first two ratios imply division into 16 parts, such division is easily obtained by folding into two equal parts, then again into two equal parts, repeatedly folding again into two equal parts to finally obtain 16 equal parts. But the relationships displayed by the last two imply division into 7 or 14 equal parts; and such divisions is not obtainable by folding in any simple manner.

Next, let us construct Haga's Second Theorem Fold. Make the fold connecting the midpoint of the top edge and one lower vertex (the right lower vertex in Fig.3.1 (c)).

We can now calculate the different lengths; see Fig.3.3. Again, be reminded that fractions referring to parts of the long side are in italics; fractions referring to the parts of the short side are in ordinary print.

Again the calculations are left to the reader. The lengths of segments calculated are as follows:

AG is $\frac{1}{9}$ of the short side — that is, short side AD is divided in the ratio 1 : 8 by G;

AI is $\frac{2}{9}$ of the long side — that is, long side AB is divided in the ratio 2 : 7 by I;

AJ is $\frac{1}{4}$ of the long side — that is, long side AB is divided in the ratio 1 : 3 by J;

AK is $\frac{2}{7}$ of the long side — that is, long side AB is divided in the ratio 2 : 5 by K.

Thus the segments other than AJ also imply odd-numbered divisions; that is, the points G, I and K can be used to divide the rectangle into 9 or 7 equal parts. Such divisions are extremely hard to obtain by folding.

3.2 Mysteries Revealed from Horizontal Folding of Copy Paper

Naturally, in a discussion of one-fold paper-folding with copy paper, one must consider both the vertical and horizontal layouts (see Fig.3.4).

First we will perform Haga's First theorem fold. Placing the $1 : \sqrt{2}$ rectangle in horizontal layout (Fig.3.4), we mark the midpoint on the top edge. We then fold the paper so that one vertex of the lower side (right vertex in the figure) coincides with the midpoint.

Unlike the folding for vertical layout, the triangular flap does not

26   Vertical and Horizontal Layouts

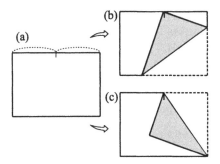

Fig. 3.4
(a) Rectangular paper in a horizontal layout,
(b) Extension of Haga's first theorem, and
(c) Extension of Haga's second theorem.

In the following figure, fractions in italics, $\frac{b}{a}$, represent the ratio when the long side is one unit.

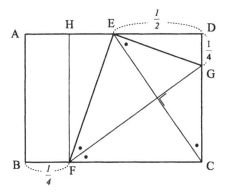

Fig. 3.5 Extension of Haga's first theorem fold of rectangle in a horizontal layout.

reach the left side (Fig.3.4(b)) and therefore initially the author had little interest in this horizontal layout. But on more careful study of his folding a revelation was awaiting him. When the paper is folded along crease FG, the segment BF remains on the bottom side and the segment DG remains on the right side. And this is the surprise: two "$\frac{1}{4}$" segments appear, BF is $\frac{1}{4}$ of the long side, and GD is $\frac{1}{4}$ of the short side of the rectangle.

The reader is invited to verify this with calculations; the auxiliary lines in Fig.3.5 may be useful. Here HF is parallel to AB.

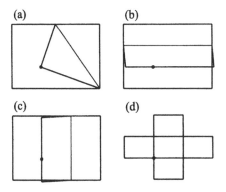

Fig. 3.6 A division point is formed in a rectangle in horizontal layout.

One other observation concerns the areas. Since HF is parallel to AB and BF is $\frac{1}{4}$ of the long side of the rectangle, the area of the rectangle ABFH is $\frac{1}{4}$ of the area of the original rectangle; the area of $\triangle$EFH is half of the area of rectangle ABFH or $\frac{1}{8}$ of the area of the original rectangle; and the area of $\triangle$DEG is $\frac{1}{8} \times \frac{1}{2}$ or $\frac{1}{16}$ of the original rectangle.

After subtracting these areas from the rectangle, half of the remainder is the area of $\triangle$EFG, or $\frac{9}{32}$ of the rectangle, roughly between $\frac{1}{4}$ and $\frac{1}{3}$ of the rectangle. This poses an interesting tidbit for the reader.

Comment. Recently while traveling on the Shinkansen Superexpress (Bullet train), I had a mysterious feeling as I performed Haga's Second Theorem Fold with $1 : \sqrt{2}$ paper. Because it is my habit to do paper folding when using public transportation, people sometimes turned their heads and cast pitying eyes on me; but because of my strong concentration at these times, their looks did not bother me.

On one particular occasion, still on Haga's Second Theorem Fold,

I was unmindful of the skeptical glances of the other passengers, and my attention was fixed on the new position of the vertex of the triangular flap. I marked this new position with a black dot (Fig.3.6). Then in an experimental frame of mind I made folds through this dot parallel to the edges of the rectangle, two more parallel to the bottom edge and equidistantly placed, and two more parallel to the side edges and equidistantly placed. It appeared that the horizontal folds divide the rectangle into three smaller rectangles of equal area (Fig.3.6(b)), and the vertical folds divide the rectangle into three smaller rectangles (Fig.3.6(c)). And a discovery dawned: the black dot seems to be a starting point for dividing the sides of the rectangle into three equal parts!

This was a big surprise for me. I have not yet been able to use this discovery in my classes, but I am sure that it will also surprise my students. "Surprise" and "mystery" are the motives for science, and tasting the emotion that accompanies such elucidations lays the ground for succeeding discoveries.

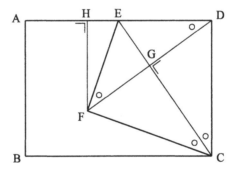

Fig. 3.7 Verification of the 3 equal parts using Haga's second theorem fold for a rectangle in horizontal layout.

Now let us prove the above result. In Fig.3.7 the intersection of DH and the crease CF is called G, and the foot of the perpendicular from F to side AD is called H.

Because the ratio of the sides of a $1:\sqrt{2}$ rectangle remains the same when the rectangle is divided crosswise into two equal parts, then the rectangle with sides CD and DE and diagonal CE is also a $1:\sqrt{2}$ rectangle. And therefore, for the sides of △EDC,

$$ED : DC : CE = 1 : \sqrt{2} : \sqrt{3}.$$

Because the following right triangles mutually share one angle other than the right angle,

$$\triangle EDC \sim \triangle EGD \sim \triangle FHD.$$

Assuming that the short side of rectangle ABCD is 1, then the long side is $\sqrt{2}$ and for △EDC, $CD = 1$, $ED = \dfrac{\sqrt{2}}{2}$, $CE = \dfrac{\sqrt{6}}{2}$.

So, from △EGD ~ △EDC, $DG = \dfrac{1}{\sqrt{3}}$. From △FHD ~ △EDC, we obtain $FH = \dfrac{2}{3}$; or FH is $\dfrac{2}{3}$ of the short side $DH = \dfrac{2\sqrt{2}}{3}$, or DH is $\dfrac{2}{3}$ of the long side.

The last two statements support the observation that F can be used to divide the rectangle both vertically and horizontally into 3 equal parts.

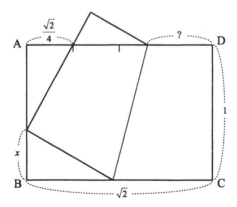

Fig. 3.8 The Third Theorem fold with A4 paper in a horizontal layout.

3.3 Using Standard Copy Paper with Haga's Third Theorem

We shall discuss both the vertical and horizontal positions of the paper.

For the horizontal layout, first locate the midpoint of the upper side and shift the lower right vertex along the left edge. You will realize that the right edge cannot meet the midpoint of the upper side.

However, if, instead of using the midpoint you take the point marking the left fourth of the upper edge and make a Third Theorem fold (that is, slide the lower right vertex along the left side), you will see that the right edge can pass through that point (Fig.3.8). It appears that the point of intersection of the fold with the upper edge cuts off a segment half the length of the shorter side; does it?

For the vertical layout - that is, the long side of the rectangle is a vertical, Third Theorem fold is possible through the midpoint of the upper edge (Fig.3.9). Look at the position of the shifted vertex on the left edge; that is, point H in Fig.3.10. It looks like it cuts off $\frac{1}{7}$ of the long side. Let us find out. Let the length of the short side of the paper be 1. Then the long side has length $\sqrt{2}$. Let HB $= x$ and BG $= y$. By the Pythagorean Theorem on $\triangle$HBG,

$$x^2 + y^2 = (1-y)^2.$$

And therefore

$$y = \frac{1-x^2}{2}. \tag{1}$$

Since $\triangle$EAH and $\triangle$HBG are similar,

$$\text{HB} : \text{BG} = \text{AE} : \text{AH}, \quad \text{or} \quad x : y = \frac{1}{2} : \left(\sqrt{2} - x\right). \tag{2}$$

Equations (1) and (2) together lead to the quadratic equation, $3x^2 - 4\sqrt{2}x + 1 = 0$. Therefore, $x = \frac{2\sqrt{2\sqrt{5}}}{3}$.

Discarding the positive sign and choosing the negative sign we obtain $x = 0.19745\ldots$.

EXTENSION OF THE HAGA'S THEOREMS TO SILVER RATIO RECTANGLES 31

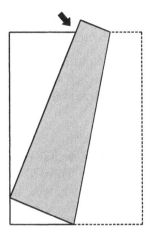

Fig. 3.9 Haga's Third Theorem fold with A4 paper in a vertical layout.

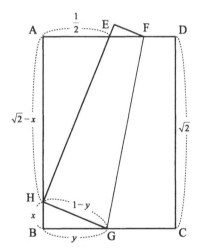

Fig. 3.10 Mathematical principles of Haga's Third Theorem fold with A4 paper in a vertical layout.

Comment. When I carried out this investigation I intuitively felt that x must be $\frac{1}{7}$ of the longer side. The value of *two sevenths* obtained in Haga's First theorem fold on a rectangular sheet of paper made me think this way. I revised the calculation by letting the length of the long side of the rectangle be 1 (and therefore the short side is $\frac{1}{\sqrt{2}}$). Still the value of x did not come out as I had hoped. So you see, some propositions may agree with intuition and others may not. But through mathematics we can judge intuitive statements very clearly as right or wrong.

There are many other discoveries that can be unearthed from carrying out Haga's First, Second and Third Theorem folds on A4 paper. I did not discuss then here, but I urge you to carry out your own investigations. And finally, try to generalize your findings to any size rectangular paper with sides in the ratio $a : b$.

Topic 4

X-LINES WITH LOTS OF SURPRISES[3]

Mathematical Ideas Related to Certain Creases Made with Respect to an Arbitrary Point

4.1 We Begin with an Arbitrary Point

In the preceding topics we started with the midpoint of an edge of a square piece of paper. This time we mark any point on the upper edge.

Many students would probably be bewildered by this instruction. They are more accustomed to following more definite instructions, not selecting any point.

The folding method is as follows:

(a) Start with a square piece of paper. Mark any point on the upper edge of the paper. While the midpoint or endpoint is acceptable, it would be more in keeping with the term "arbitrary" to pick some other point.

(b) Place one lower vertex on the selected point and make a firm crease similar to the First Theorem fold (Fig.4.1(b)). Press the whole crease repeatedly with the finger nail to make it clear and distinct.

(c) Unfold (Fig.4.1(c)). Make sure that the fold is distinct.

(d) Now place the other lower vertex on the selected point and

---

[3]These surprises come from the mathematical reasoning arise through paper folding. Through those surprises, we can find the invariant and it initiates us to inquire much more paper folding as mathematical science activity.

make another firm, distinct crease (Fig.4.1(d)).
(e)   Unfold again (Fig.4.1(e)). Do you see the two creases?

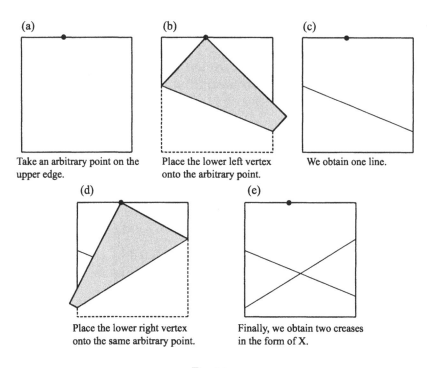

Fig. 4.1

The folding procedure described above differs from ordinary origami in at least three aspects. First, in ordinary origami the procedure involves folding, folding again, and so forth, with little or no unfolding in between. In the present procedure there is unfolding after every step. Second, in the present folding what is important are the folds, not the whole figure. It is for this reason that the creases should be distinguishable. And third, unlike in origami there is no animal or toy or work of art that comes out of the present folding procedure.

4.2 Revelations Concerning the Points of Intersection

We shall call the pair of creases obtained above as **X-creases**. With other pieces of paper make other X-creases with other starting points. Use different pieces of paper; if just one piece is used, the many creases would make things messy and judgments would be difficult. Besides, you will use the different pieces of paper later. For each piece of paper mark the starting point with a pen or pencil.

Different X-creases are obtained by different starting points, and therefore the position of the point of intersection may vary.

Take one piece of paper. Make a vertical book fold to obtain the vertical midline of the square. Do likewise with your other X-creases. Observe that, regardless of the starting point, the intersection falls on the midline (Fig.4.2).

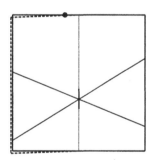

Fig. 4.2 By making a vertical book fold, it is confirmed that the intersection falls on the midline.

We state our observation:

> (1) The points of intersection of the X-creases fall on the vertical midline.

A mathematical argument for observation (1) shall be discussed later.

Another matter to be studied is the vertical position of the intersections. From your folding you can see that the intersections lie a little below the center of the square.

36   *Mathematical Ideas Related to Certain Creases*

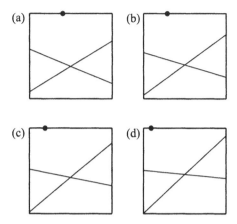

Fig. 4.3   The intersection of the X-creases changes when the position of the arbitrary point is changed.

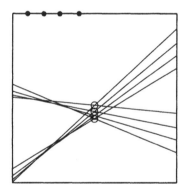

Fig. 4.4   The position of the intersection varies up and down when A, B, C, and D of Fig.4.3 are brought together.

Pile up the pieces of paper which you used to make X-creases, and hold the pile up to the light. You will see that the points of intersection seem to vary up and down along the midline, although within a small range (Fig.4.4).

We state our observation:

(2)   The points of intersection of the X-creases lie along the midline and lie below the center of the square within a certain range.

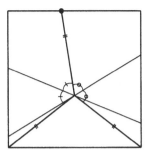

Fig. 4.5 The arbitrary point, lower right vertex and lower left vertex are joined to the intersection point.

Again observation (2) needs to be mathematically supported, but we shall take care of this later. We describe other interesting findings.

Select a starting point and the corresponding X-creases. Draw a line from the intersection point to the starting point. Also draw lines from the intersection point to the lower vertices of the square (Fig.4.5) (instead of drawing it is possible to make creases for these spokes-like lines if you prefer). Then fold along an X-crease and hold the paper up to the light. It comes out that two of the spokes coincide. Repeat with the other X-crease. It appears that the third spoke also has the same length.

We state our observation:

> (3) The distances from the point of intersection to the starting point and to each of the lower vertices are equal.

4.3 The Center of the Circumcircle!

We shall now demonstrate observations (1) and (3). Refer to Fig.4.6.

First we prove observation (1).

Draw straight lines connecting the starting point to each of the lower vertices; that is, EB and EC. This forms triangle EBC. Fold again along one of the X-creases and hold the paper up to the light. You will see that the line you just drew is folded into two equal parts, one part lying on the other.

That is, the X-crease is the perpendicular bisector of one of the lines you just drew. Likewise, the other X-crease is the perpendicular bisector of the other line.

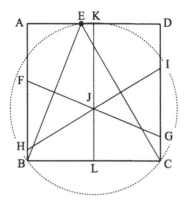

Fig. 4.6 The intersection point, J, of X-creases is the circumcenter of △EBC (the center of circumcircle).

Recall that in a triangle the perpendicular bisectors of the sides intersect at a common point. Since J is the intersection of the perpendicular bisectors of sides EB and EC, then J should also lie on the perpendicular bisector of side BC - that is, the vertical midline. Thus statement (1) has been proved.

Recall further that the intersection of the perpendicular bisectors of the sides of a triangle is equidistant from the vertices; that is, it is the center of the circle containing the vertices of the triangle. In other words, J is the center of the circumcircle of △EBC. This proves statement (3).

Certainly these paper folding experiences could arouse in the students greater interest in what they learned in their traditional mathematics course.

4.4   How Does the Vertical Position of the Point of Intersection Vary?

We now discuss observation (2).

Looking back at Fig.4.4 it appears that the vertical range of variation of the intersections is rather short. And furthermore, it appears that the lowest point of intersection is obtained when the starting point is the midpoint of the side. The situation where the starting point is the midpoint is shown in Fig.4.7, and the position of the intersection point can be calculated by the Pythagorean relation and the proportionality of corresponding sides of similar triangles. Letting the sides of the square be of length 1, the point of intersection is found to be $\frac{3}{8}$ from the lower edge.

The highest point of intersection is obtained when the starting point is an endpoint of the edge (Fig.4.10). Here the point of intersection coincides with the center of the square, that is, $\frac{1}{2}$ from the lower edge.

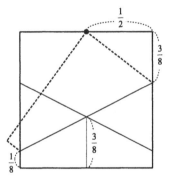

Fig. 4.7 The position of the intersection of X-creases by the First Theorem fold.

These values may be obtained more precisely by representing the X-creases as equations and solving the system. Let the x and y-axes be two sides of the square and let a be the distance of the starting point from the y-axis, as in Fig.4.8.

In order to obtain the equation of the creases we need to express the coordinates of the endpoints in terms of a. The discussion for Fig.1.7 of Topic 1 may be helpful.

Equation (i) refers to the crease obtained by folding the lower left vertex onto the starting point; equation (ii) refers to the crease obtained

40 Mathematical Ideas Related to Certain Creases

by folding the lower right vertex onto the starting point.

$$y = -ax + \frac{a^2}{2} + \frac{1}{2} \qquad \text{(i)}$$

$$y = (1 - a)x + \frac{a^2}{2}. \qquad \text{(ii)}$$

Solving simultaneously to obtain the point of intersection, we obtain

$$-ax + \frac{a^2}{2} + \frac{1}{2} = (1 - a)x + \frac{a^2}{2}.$$

Therefore $x = \frac{1}{2}$. Substitute this value of x in equation (ii) to obtain

$$y = \frac{1}{2}\left(a - \frac{1}{2}\right)^2 + \frac{3}{8}.$$

Since $0 \leq a \leq 1$, then $\frac{3}{8} \leq y \leq \frac{1}{2}$.

Thus the range of variation for the point of intersection is $\frac{1}{2} - \frac{3}{8} = \frac{1}{8}$, which proves the result of Figs. 4.7 and 4.8.

Incidentally, the value of x obtained here confirms our earlier statement that the points of intersection lie on the midline.

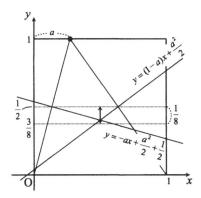

Fig. 4.8 Calculating the range of variation for the point of intersection by thinking X-creases as a graph.

Comment. The above solution was presented by Mrs. Yoko Takamoto, a teacher of the Senior High School attached to Toshimagaoka Women's Educational Institution, while a trainee at Tsukuba University's Extension course entitled "Origami and Education".

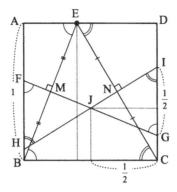

Fig. 4.9 The length FH (= IG) is an interesting fact.

4.5 Wonders Still Continue

Let us now study the lengths of the segments FH and IG, the segments on the sides of the square intercepted by the X-creases (Fig.4.9). These are sides of triangles FJH and IJG respectively. Since J lies on the midline then FJ = JG and HJ = JI. Also, since they are vertical angles, ∠FJH = ∠GJI. Therefore △FJH ≡ △GJI, so FH = IG.

Using the pieces of square paper that you folded for different initial points match the lengths of these segments. It comes out that they are identical in length, even as the position of E is allowed to change. How fascinating!

We state this observation:

> (4) The segments on the sides of the square intercepted by the X-creases have a fixed length, regardless of the chosen position of the initial point.

Be reminded that this statement is the result of comparing with pieces of paper, so at this time it would be more accurate to add the phrase "seems to be", or "highly possible". But for a while let us leave it as is.

If all segments FG and IG, are equal, then what is their length? Since the selection of the initial point does not affect the length, we can select a particular one. Select an endpoint as the initial point as in

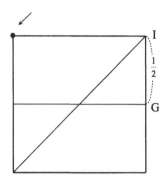

Fig. 4.10 When the endpoint is chosen as the arbitrary point, it is obvious $IG = \frac{1}{2}$.

Fig.4.10. Judging from the figure one can easily understand that the segment FH is one half of the side of the square.

Therefore on the assumption of (4), we have arrived at:

> (5)   The lengths of the segments intercepted on the sides of the square by the X-creases are $\frac{1}{2}$ of the side of the square.

4.6 Solving the Riddle of "$\frac{1}{2}$"

Observation (4) still needs to be proven. We use Fig.4.9. The statement can be proved if it can be shown that triangles BEC and FJH are similar, and that the height of △FJH is half that of △BEC.

With this in mind, first look at △BAE and △BMF. Since both triangles are right and have a common angle, they are similar. Therefore ∠BEA = ∠BFM (or HFJ).

Likewise, on the right side, look at △CED and △CIN. They are also similar, so ∠CED = ∠CIN (or GIJ).

Finally look at triangles BEC and FJH. Since ∠BEA and ∠EBC are alternate interior angles of parallel edges, they are equal. Therefore ∠EBC = ∠JFH. Likewise, ∠ECB = ∠JHF. Therefore triangles BEC and FJH are similar. But the height of the larger triangle is 1 and the height of the smaller is $\frac{1}{2}$ (since J is on the midline), therefore the ratio of their heights, as also corresponding sides is 1 : 2. In particular, FH is $\frac{1}{2}$ of a side.

4.7 Another Wonder

If X-creases are constructed not only for squares but also for rectangles, would statements (1) to (5) hold? Let us study this matter by experimenting with different shapes of rectangles. Include a half-square (that is, cut a square along a midline), and a size A4 paper. For the half-square the ratio of the sides is 1 : 2, while for the A4 paper the ratio of the sides is $1 : \sqrt{2}$.

The position of the rectangle should be such that the shorter sides are the upper and lower edges; otherwise in the folding the creases will be forced off the paper through the lower edge instead of through the left and right edges.

The result of your experimenting should lead you to believe that statements (1), (2), (3) and (4) apply to all rectangles. The mathematical proof is left to the reader.

From your experiments, notice that as the rectangles become narrower and longer the X-creases become flatter, and the intercepts on the sides become shorter. Therefore, for rectangles statement (5) will need to be revised. Fig.4.11 helps us visualize the problem for the general rectangle with dimensions a and b. Calculating the length of FH in terms of a and b is straightforward and is left of the reader. The table following summarizes the results for different sizes of rectangles.

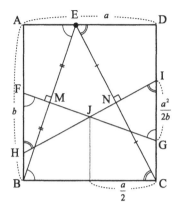

Fig. 4.11 Generalization of the mathematical principles of X-creases.

Size of Paper	Ratio of short side to long side	Range of variation of y-coordinate of intersection point
Square	1	$\frac{1}{8}$ of long side
A4 paper	$\frac{1}{\sqrt{2}}$	$\frac{1}{16}$ of long side
"Half-square"	$\frac{1}{2}$	$\frac{1}{32}$ of long side

The outcomes in the last column are fascinating; do you see the geometric sequence?

Topic 5

"INTRASQUARES" AND "EXTRASQUARES"

Drawing the Map of the Distribution of the Position of the Reference Vertex According to the Shape of the Flap Formed by Folding a Piece of Paper Once

Most of the subject matter that has been taken up so far had as prerequisite knowledges the Pythagorean Theorem, quadratic equations, incenter, circumcenter, and excenter of triangles, and therefore are more suited to third year middle school (9th grade) students and above. But there have been requests for topics within the grasp of first and second year middle school (7th and 8th grade) students, so here is a topic which could foster logical thinking that does not rely on too many theorems and equations.

Comment on Teaching. I have tried using this lesson in lectures to third year middle school (9th grade) students and in popular lectures, and even to fourth to sixth grade elementary school pupils and first and second year middle school (7th and 8th grade) students. While it is an effort to cram the whole lecture into a 45 to 50-minute period, it is possible to divide the lecture into two parts: *Inner Subdivision* and *Outer Subdivision* (these terms shall be explained later); or conduct the lesson during more flexible time slots, or do it as a group activity.

Here now is the lesson, presented as a sequence of instructions to a group of learners.

5.1 Do Not Fold Exactly into Halves

An origami square of any size or color will do. A square cut from ordinary bond paper will do, but it would be easier for the teacher to give instructions and the pupils to make observations if the two sides can be distinguished - say one side is colored. Distribute four or five sheets for each pupil.

In making an ordinary origami object (e.g., a crane), the first fold is either the book fold or the diagonal fold, each one dividing the square into two equal parts. But for the present topic, the square is not divided into two equal parts. Make sure the pupils are aware of this fact before giving the following instructions.

"Put one sheet on the table. With the colored side facing down and the white side up, fold only once. Do not fold the paper into two equal parts. In other words, you are not to put a corner or an edge exactly on another. You are to make a *random* fold. You may fold the paper at any place or in any direction, but fold it neatly and carefully."

At first the pupils might be hesitant and may look around for encouragement. Remind them, "Fold only once."

After the pupils are done folding once, ask them to raise their work so everyone else can see them. There will be a variety of folds. Some may make a short fold that cuts off only a small portion of the paper (Fig.5.1(e)), while others may make a long fold to cut off a larger portion (Fig.5.1(d) and (k)).

5.2 What Kind of Polygons Can You Get?

"Some of you made short folds while others made long ones, but the length of the fold does not matter. What we are interested in is the shape of the colored flap that you see after you folded the paper. If it has three corners then it is a triangle; if it has four then it is a quadrilateral."

The pupils should find it easy to identify the shapes. Ask those with triangles to raise their pieces of paper; then also those with quadrilat-

"INTRASQUARES" AND "EXTRASQUARES"

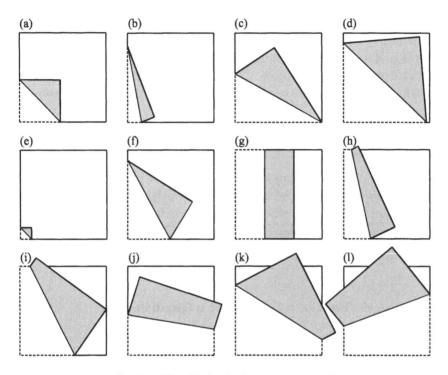

Fig. 5.1 What kinds of polygons can you get?

erals. Ask if other shapes were obtained. Some pupils may come up with five-sided figures (pentagons); if there are, tell them, "Good. We will study that later, so please save your work." If some report six or more corners, they could be looking at the whole paper and not just the colored portion, so call their attention to this.

Figure 5.1 shows examples of folds that result in triangular flaps ((a) to (f)) and quadrilateral flaps ((g) to (l)).

Comment. In my lecture at a third year middle school (9th grade) for girls in Tokyo, of the 123 pupils the number who came up with triangles was almost the same as the number who came up with quadrilaterals. In a special lecture given to 150 pupils in the higher grades at an elementary school in Yamagata Prefecture, two fifths had triangles and about three fifths had quadrilaterals. In a lecture at a middle school in Ibaraki Prefecture participated in by 80 first year middle school pupils and 40 guardians, three fifths formed triangles. In an origamics workshop mainly for teachers but with about 40 nonteachers attending, two thirds had triangles while only one third had quadrilaterals. One came up with a five-sided polygon. The size of the paper and the manner by which the folding was demonstrated may account for the differences mentioned.

Let us proceed to the next step.

Ask the pupils to bring out another piece of paper. Then instruct them: "Now those who had triangles previously, make a fold to form a quadrilateral flap; those who obtained quadrilaterals previously make a fold to form a triangular flap." This way each pupil would have the two shapes.

5.3 How do You Get a Triangle or a Quadrilateral?

Ask the pupil to examine the two kinds of flaps that they made and compare the manner by which they were folded. If, by just using their own pieces of folded paper as examples they find it difficult to figure out conditions by which a triangle or quadrilateral is formed, make different folded flaps as shown in Fig.5.1. Pupils tend to think that there is only one "correct answer", so remind them there are many different valid ideas.

After some time, expressions on some faces would indicate that they have somehow figured out something. Ask them to tell what they found out.

We can expect to hear different ideas from them, here are some possibilities.

Mr. A: Open the folded paper and examine the line of the fold. If the line connects two adjacent sides of the original origami square, then the flap is a triangle; if the line connects two opposite sides, then the flap is a quadrilateral.

Ms. B: If in folding only one vertex of the original square moves, then the resulting flap is a triangle; if two vertices move, then the flap is a quadrilateral.

Mr. C: The shape of the flap depends on the position of the moved vertex on the origami paper.

Ms. D: If the colored portion is contained wholly in the origami square, then it is a triangle; if a part of it is outside the square then the flap is a quadrilateral.

Mr. E: If the line of the fold passes through any of the vertices of the square, then surely the resulting flap is a triangle. If the line of fold does not pass through any vertex then the flap could be either a triangle or a quadrilateral.

Ms. F: If the line of fold is shorter than a side of the square, then the flap is a triangle; if longer, then we could have either of the two.

These are just some of the many possible discoveries that the pupils may make. Tackle each one of them by asking the others what they think of the comment. For example, one may say that Mr. A's idea is not applicable when the line of fold hits a vertex. As for Ms. D's idea, a counterexample can be found by folding parallel to a side (Fig.5.1(g)). - that is, a quadrilateral may be formed even if the colored flap falls entirely on the paper square. It is important that when leading the discussion, one should not judge right away that an idea is "right" or "wrong" For example, one should not say that Mr. X's answer is correct or Ms. Y's is wrong. Each person has discovered something, although an idea may work only in certain situations. The teacher can clarify by identifying the conditions or cases where the idea works. The ideas given by the pupils should help the teacher in deciding what to examine or explore next.

5.4 Now to Making a Map

We continue looking for conditions that lead to the formation of a triangular or a quadrilateral flap. One way is to pursue the line of thinking of one of the pupils. Ms. B and Mr. C focused on the movement of one vertex; let us pursue Mr. C's idea. Use a new sheet of square paper. Mark at random about ten points on the paper. Select one vertex of the paper square to be the reference vertex (in Fig.5.2 the reference vertex is the lower left vertex, marked ○). Then carefully move it to a marked point and determine the shape of the flap formed. Write "3" or "4" beside the point or color it, to indicate the number of the vertices of the polygonal flap.

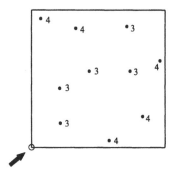

Fig. 5.2 Investigating what kind of polygons we can get by plotting points randomly.

Do this for the other marked points. As more and more of the random points are labeled, a map begins to be formed. Notice how the "3's" cluster in certain areas of the paper square; so also the "4's". If one thinks of the square as a village divided into lots, then the clusters of "3's" may be called "Lot 3" while the clusters of "4's" may be called "Lot 4".

In order to better identify the boundaries of Lot 3 and Lot 4, let us repeat the experiment, but this time select points more systematically (if your paper is messy, use a new sheet). Using a ruler draw equally-spaced vertical and horizontal lines to make a grid. For a piece of square paper with side 15 centimeters an interval of 1 centimeter between lines will work fine. A sheet from a ruled notebook or ruled pad paper, cut into a square, may be used (see Fig.5.3). Again select a reference vertex (marked ○ in Fig.5.3), and move it to each of the intersection points of the grid. As before, label each point as "3" or "4" to indicate the shape of the flap obtained.

"INTRASQUARES" AND "EXTRASQUARES"

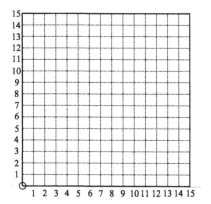

Fig. 5.3 Setting the coordinates, and investigating in groups.

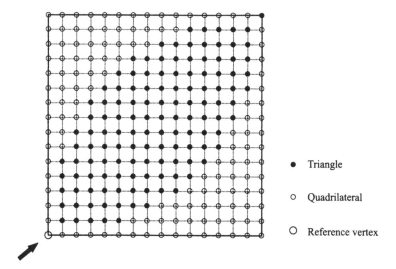

• Triangle

○ Quadrilateral

○ Reference vertex

Fig. 5.4 The distribution area gradually appears.

The points on the sides of the squares should also be examined and labeled. Thus, for a 15 cm square with horizontal and vertical lines at 1 centimeter intervals, a total of $(15+1)^2 - 1$ or 255 points must be examined. As it would take time for any one pupil working alone to check all the points, it is recommended that they work in groups and

52   *Drawing the Map of the Distribution of the Position*

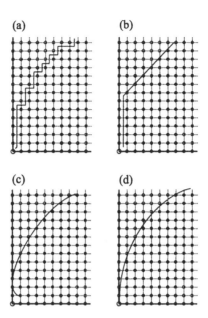

Fig. 5.5   It is necessary to examine the line near to the boundary in detail. (It is easy for students to draw the zigzag line (a).)

let the members of a group divide the work among themselves.

Figure 5.4 shows all 255 points labeled. However, in the actual classroom it is likely that, even before all 255 points are worked on, the pupils will recognize a pattern and identify the regions of Lot 3 and Lot 5. Let the pupils look for the boundaries by checking more points, including some not on the grid. Countercheck the children's work by folding and checking the resulting flaps for different points. Take note that in drawing the boundary many students tend to use straight lines as in Fig.5.5(a) and (b); if they do present such a boundary, remind them to examine more finely the regions near the boundaries.

Thus in the manner just described the boundaries can be identified, checked, redrawn if necessary, and once again verified. This is an ideal flow of the pupils' activity; it is therefore very important that the lecturer give subtle hints to lead the discussion in the desired direction.

5.5 This is the "Scientific Method"

The activity of making a map described earlier might seem to be just a casual one; actually there was a method to the undertaking. The activity is really an *experiment* based on the *scientific method*.

One becomes conscious of a certain phenomenon, a question arises ("in folding a piece of paper, how do we end up with triangles or quadrilaterals?"), then one sets his mind to understanding the phenomenon (the beginning of science). First hypotheses or conjectures are made (ideas of Mr. A to Ms. F). Based on one or more conjectures a preliminary experiment is conducted (studying random points). Then an organized plan is conceived, and preparations for a more systematic experiment are made (making grids). The experiment is then performed (reference vertex is chosen and made to point to the grid points and the paper is folded to identify the figure formed by the flap) and data is gathered, the experiment is modified based on the gathered data (thorough examination of the boundaries). The conjecture is redefined (properties of the boundaries are considered and a map is drawn), and the results are verified (more marked points and paper folding to confirm). If at one step something is found not to agree or fit with the conjecture, a new conjecture is made, the experiment is modified and the steps are repeated.

5.6 Completing the Map

The resulting map is shown in Fig.5.6. If the lower left vertex is selected as reference vertex (marked by o in Fig.5.6) and is moved to a point anywhere inside the region enclosed by the arcs, a triangle flap is formed. Pointing the reference vertex onto the sides of the square would also yield triangles. If the reference vertex is moved outside the said region the resulting flap would be a quadrilateral. The two regions may then be called the Triangle Lot or Region and the Quadrilateral Lot or Region respectively.

The map of Fig.5.6 shows two circular arcs with centers the vertices adjacent to the reference vertex and radii the length of the side of the square. We arrived at this figure through experimenting, but this

54   *Drawing the Map of the Distribution of the Position*

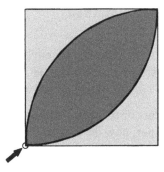

Fig. 5.6   The eyeball shape at the center is the triangle region whereas the outsides are the quadrilaterals region.

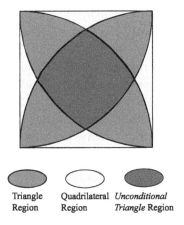

Fig. 5.7   A definite region would produce triangle no matter which vertex is used.

actually results from the ideas of pupils Mr. A and Ms. F. In other words, the boundary between the Triangle and Quadrilateral regions is the path traced by the reference vertex as it moves, with one end of the line of the fold fixed at one of the vertices adjacent to the reference vertex. This condition is the same if the vertex diagonally opposite is selected as the reference vertex. Thus, whichever vertex is moved, the central region in Fig.5.7 will always be a triangle region; or we could say the *Unconditional Triangle* Region.

5.7 We Must Also Make the Map of the Outer Subdivision

Let us go back to Fig.5.1. Here the reference vertex is the lower left vertex of the original square. Note that in figures a to j the reference vertex points to somewhere in the square, and hence Figs.5.5 and 5.6 apply. But look at the last two figures k and l. In both a quadrilateral is formed, but the reference vertex does not point to somewhere in the square; rather it points to somewhere outside the square (see also Fig.5.8 (a) and (b)). In other words, the reference vertex has gone to the *Outer Subdivision*. In contrast, the *Inner Subdivision* refers to points on the paper - that is inside or on the square. Thus, in examining the path of the reference vertex that would lead to quadrilaterals, it is not enough to consider just the sheet of origami paper; the area outside must also be examined.

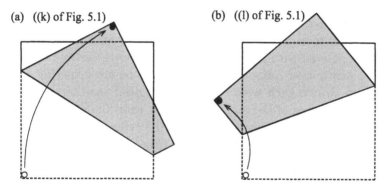

Fig. 5.8 Quadrilaterals could be obtained even if the reference vertex falls outside the square paper.

It was mentioned in section 5.2 that there was one person that came up with a pentagon flap. Figure 5.9 shows how that person folded his paper. So in the Outer Subdivision, a Pentagon Region actually exists. Extending the observation of Ms. B, we could say, "If the three vertices of the square move then the resulting flap is a five-sided figure or pentagon."

In order to obtain the map of the Pentagon Region of the Outer Subdivision we will need a different, larger sheet of paper. Place the origami square somewhere in the middle of the large sheet and keep

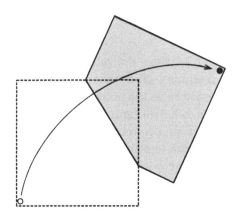

Fig. 5.9 A pentagon could be obtained depending on the placement of the reference vertex.

its position fixed. Only the reference vertex is allowed to move; after the move the reference vertex must return to its original position. One can see that even with this condition the reference vertex can actually move to points over quite a large region. To be more specific, if three circles are drawn each with center the original position of the three remaining vertices of the square and with radii the distance of these vertices from the reference vertex, then the reference vertex can point to anywhere in or on these three circles (Fig.5.10).

To make the map of the Outer Subdivision, draw the said circles and by folding the origami paper, examine separately the regions enclosed by the intersecting circles. If the pupils had started with a square 15 cm on a side, the region to be examined would span about 50 cm in four directions; so for the sake of convenience start with a smaller origami paper; or work only with a portion of the whole region.

The resulting map is shown in Fig.5.10. To highlight the symmetry of the map, the drawing was rotated so that the diagonal of the original square is vertical with the reference point at the lower endpoint (indicated by an arrow)

I was surprised and elated that the simple of instruction of "folding the origami paper just once" would yield such a figure; I had the figure neatly enlarged and hung on the wall. "Patterns such as this could arise in the unconventional world of origamics," I would chuckle. And

"INTRASQUARES" AND "EXTRASQUARES"

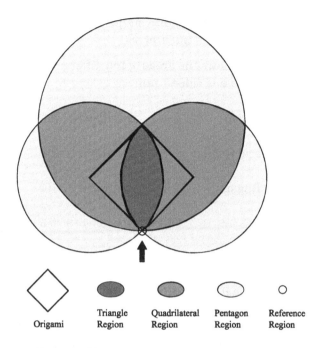

Fig. 5.10 This shape is obtained with only one fold.

when others ask me what the figure is, I would hide my pride and solemnly say, "Oh, just an abstract illustration of an insect." I enjoyed keeping the figure a mystery.

5.8 Let Us Calculate Areas

Up to this point no equations or geometric symbols were used, so that people might feel that the whole activity is quite "un-mathematical". In a broader sense it might be better to term it a scientific one. In this section we bring out a mathematical aspect of the activity.

By using Fig.5.10 it can be verified that the region where the three circles intersect is the Triangle Region (Lot 3), the region where exactly two circles intersect is the Quadrilateral Region (Lot 4), and the remaining region is the Pentagon Region (Lot 5).

Let us compare the areas. We will not discuss in detail the method

Drawing the Map of the Distribution of the Position

for calculating the areas; this is quite simple, and I think the learners will be happy to work on it by themselves.

The following table shows the areas of the different regions; assume that the original square has sides 1 unit.

Area of Triangle Region	$\frac{\pi}{2} - 1$	(or about 0.5708)	6.89%
Area of Quadrilateral Region	π	(or about 3.1416)	37.93%
Area of Pentagon Region	$\frac{\pi}{2} + 3$	(or about 4.5708)	55.18%
Total area	$2\pi + 2$	(or about 8.2832)	100.00%

Note that the Pentagon Region has the greatest area; theoretically speaking the probability that a pentagonal flap is obtained from a random folding is more that 50%. However, in reality very few persons would come up with pentagons. It is more likely that a person folding randomly would come up with a triangular flap; yet the area of the Triangular Region is smallest. Can this seeming paradox be explained? This might be an interesting study in psychology.

Topic 6

A PETAL PATTERN FROM HEXAGONS?

Making a Map of the Distribution of Points According to the Kinds of Polygonal Figures Generated by Concentrating the Vertices at a Point

6.1 The Origamics Logo

Comment. I first showed the flower-like pattern with four petals (Fig.6.6) in 1994 at the session of the Second International Congress of Origami Science and Scientific Origami held in Japan, then again in 1995 at the lecture of the Annual Meeting of the Japan Mathematical Education Society in Tokyo. Some newspapers such as the Asahi (October 31, 1995) and the Newsletter of the Society picked up the pattern. Since then the petal pattern has become popular among mathematics teachers, origami students and hobbyists; and it became the first logo of origamics.

We begin the folding procedure by marking an arbitrary point on a square piece of paper. By folding so that all the vertices of the square fall on this point, a polygon is formed. We shall see that the kind of polygon that emerges - e.g., quadrilateral, hexagon, etc. - depends on the position of the original point. Therefore it is possible to draw a map on the paper showing the distribution of the kinds of polygon that come up.

Students should find this topic interesting because the result becomes unusually attractive, especially if color is used.

60   Making a Map of the Distribution of Points

6.2 Folding a Piece of Paper by Concentrating the Four Vertices at One Point

Two folding procedures will be presented here. The first is simpler, but the second gives more impressive polygons. Have ready several pieces of square paper - say four or five.

Folding Procedure 1.

Figure 6.1 illustrates the folding procedure. Different starting points result in different creases, so your creases may look slightly different from the figure.

1.1   Select any point on the paper and mark it. Any point will do, but to follow the diagrams, for now select a point in the upper left quadrant of the paper as in Fig.6.1(a). Call this point the *point of concentration*, or just *concentration point*.

1.2   Select a vertex of the square paper. Fold so that this vertex falls on the selected point. Make a firm crease by holding the paper steady and rubbing the crease with something hard, say your fingernail (Fig.6.1(b)). Unfold (Fig.6.1(c)).

1.3   Repeat step 1.3 for the other vertices of the square paper.

Four creases result (Fig.6.1(d)). And the arbitrarily marked point is enclosed in a polygon made by the folds and one or more edges (Fig.6.1(e)).

Look at the work of others in the class. You will probably see that not all the polygons are of the same type. Take another piece of square paper and repeat the folding procedure with a different starting point. If the first figure you obtained was a hexagon, select a point farther from the diagonal lines of the square and nearer an edge. Otherwise select a point nearer a diagonal line. Fold as before.

Folding Procedure 2.

2.1   A second folding procedure is to carry out the steps in the first folding procedure without unfolding at each step - place one vertex on the starting point, then another vertex, then another

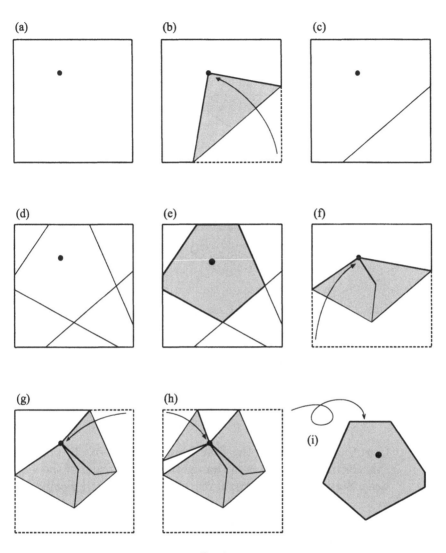

Fig. 6.1

vertex, then the last (Fig.6.1(f) to (h)). For a neat polygon, turn the figure over (Fig.6.1(i)). It could happen that when a vertex is placed on the concentration point one of the free vertices would move from its original position as in Fig.6.3(a). If the folding is continued the resulting polygon would be different

62   Making a Map of the Distribution of Points

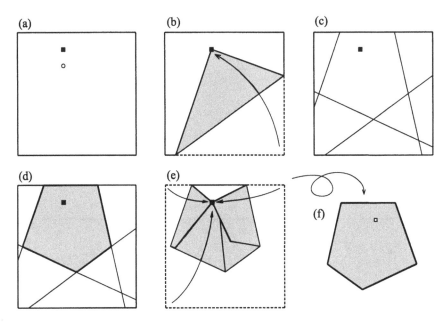

Fig. 6.2   ○: The original arbitrary point of Fig.6.1   ■: The new arbitrary point in this time
A pentagon instead of a hexagon is formed with just a slight change in the position of the arbitrary point.

from that obtained by the first folding procedure. To prevent this shifting, one way is to change the folding order of the vertices. Start with the vertex farthest from the starting point.

This change of order works sometimes, but not always. For instance, if the arbitrary point is near a midpoint of an edge, no folding order would prevent shifting of a free vertex. In such a case, go to step 2.2.

2.2 Before firming up the crease when you position a vertex, place the vertex and the next simultaneously on the starting point, and press the paper. If there is any protruding part of the square, tuck it in, then make a firm crease (Fig.6.3(b) to (d)).

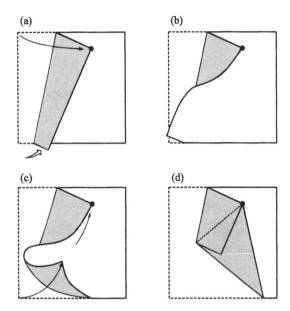

Fig. 6.3 The processing method when other vertices move is to slacken the paper and make the vertex onto the arbitrary point.

6.3 Remarks on Polygonal Figures of Type n

Polygons may be classified according to the number of sides or number of angles, regardless of the lengths of the sides or the sizes of the angles - that is, independently of the differences in shape. The different types according to the number of sides (or angles) may be described as *polygons of type n*, where n is the number of sides (or angles). Thus we describe pentagons as polygons of type 5, and hexagons as polygons of type 6.

In the preceding sections we saw how n depends on the position of the point of concentration. That is, n is uniquely determined by the position of the point of concentration. On the square piece of paper we saw that some points of concentration yield $n = 6$, others $n = 5$. So the points on the paper may also be classified according to the related value of n. It is likely that the set of points of the same type occupies a certain definite area on the paper, and the set of points of a different type a different area.

Making a Map of the Distribution of Points

We now study how the points on the paper are distributed according to polygonal type, and draw the map of the distribution.

6.4 An Approach to the Problem Using Group Study

In this section a teaching approach is described which encourages experimentation and investigation. It was carried out on a class of thirty students in the third grade of a Middle School (ninth grade). The topic was not part of the regular course of the study, and the students worked on the problem for two school periods (100 minutes).

The class was divided into six groups of five persons according to the alphabetical order in the class roll book, and the study was undertaken by each group. At first no hint was given except for the following two suggestions: (a) an experimental approach may be used, and a piece of paper may be used repeatedly; (b) an efficient way of collecting experimental data may be carried out by letting the group members share in the work. And the groups were instructed to plan their work strategy according to the order of steps below.

(Working strategy)

Group Activity

- First planning meeting. The group members plan how to approach the problem efficiently, how to divide the work among them, and to plan a time schedule.
  The group members would then do exploratory paper folding work and organize their results.

- Second planning meeting. The group members discuss the results of their work so far, looking for patterns in order to set up a hypothesis. Then they plan how to verify their hypothesis, and how to do the paper folding more systematically.
  The group members would then proceed to do their paper folding in their systematic way, organize their results, and describe the distribution map.

Individual Activity

> • Each member would draw the distribution map resulting from their observations in as nice a format as possible. Use of color was encouraged. Finally they would try to analyze why the resulting distribution pattern does emerge.

From the outcomes of the planning meetings three different work patterns were identified. In the list of patterns below the numbers in parentheses indicate the number of groups who exhibited the respective pattern.

Random approach (1)
   The square paper was roughly zoned — e.g., central part, peripheral part, part near a vertex — and the group members shared in the investigative work.
Theory approach (2)
   The group tried to figure out a condition for a fold to produce a corner of the figure.
Coordinates approach (3)
   The group members set up coordinate axes on the paper then tackled different areas delineated by the axes, and therefore the members shared in the work. One group divided the paper into left part and right part relative to the y-axis.

6.5 Reducing the Work of Paper Folding; One Eighth of the Square Will Do

For a while the sixth group appeared at a standstill. They did not know what plan of action to take. After a while the instructor gave a hint:

> "Consider setting up coordinate axes. Select some points on the coordinate plane and find the polygon type for each selected point. Do some paper folding first, then pause, examine for a pattern, and decide if it is really necessary to fold for all the points on the piece of paper."

With this hint the group members were able to get to work. They drew horizontal and vertical lines through the center of the square for the axes. They guessed that working on the first quadrant may enable them to obtain the map for the other quadrants. After a while they decided to divide the first quadrant into two smaller triangles by a diagonal line through the center of the square (Fig.6.4(a)). Finally they investigated the lattice points of one of the triangles (Fig.6.4(b)). Then they extended the pattern to the other triangle, and finally to the other quadrants. Thus in the course of their working they recognized that the triangle pattern would be symmetric with respect to the diagonal line, and the quadrant pattern would be symmetric with respect to the axes and the center. With this hint the sixth group successfully showed that their ideas saved themselves some effort.

In the end, five of the groups reached the four-petals pattern of Fig.6.6. The random group could only roughly demarcate the pentagonal and hexagonal areas; the demarcation lines were not well defined.

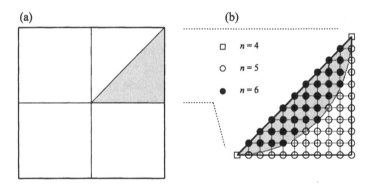

Fig. 6.4

6.6 Why Does the Petal Pattern Appear?

The petal pattern may be obtained by repeatedly folding the paper for many points, thus empirically pursuing the result. But, as we have seen in the previous section, with some analyzing the paper folding work can be reduced.

The pattern consists of four semicircles intersecting at the center of the paper. Their centers are the midpoints of the sides, and their radii half the length of a side. The overlapping areas form the four petals. If a point of concentration is selected inside two circles (that is, inside the petals), a hexagon is obtained; if a concentration point is selected inside one circle only (that is, outside the petals) a pentagon is obtained. For this reason we may call the area inside the petals as the *hexagon region*, and the part outside the *pentagon region*.

There is still the case of concentration points on the border between the two zones — that is, on the rim of the petals — and concentration points at the vertices. If any of the vertices is chosen a square is formed with area one fourth that of the square. And if one picks the center of the square as point of concentration the resulting figure is a square as well, with area half the size of the square paper (Fig.6.5). All other points on the outline of the petals induce pentagons.

This exhausts all points on the paper. Therefore we conclude that, by the vertex-concentration folding method, three types of polygonal figures are produced: $n = 4, 5$ and 6. The distribution of the points of concentration according to polygonal types is shown in Fig.6.6.

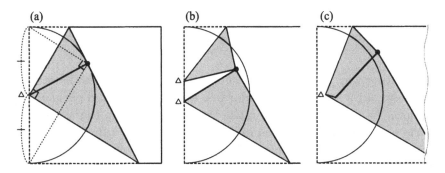

Fig. 6.5 The position of the arbitrary point determines the formation of either a pentagon or hexagon.

But why does such a pattern appear? With a clean paper square fold down loosely and hold together two adjacent vertices. Observe how the common side shifts as you move to different concentration points (Fig.6.5). Part of the common side may remain exposed as in Fig.6.5(a), and therefore two corners of the final polygon appear. Or no part of the

68   Making a Map of the Distribution of Points

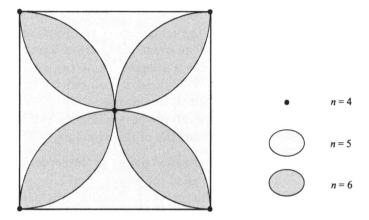

Fig. 6.6   The distribution chart becomes a beautiful petal pattern.

common side remains as in Fig.6.5(b) and (c), and therefore only one corner of the final polygon emerges. The critical point at which no part of the common side appears is that where the sides of the upper triangular flap and the lower triangular flap are equal, or one half of the side of the square. Thus the locus of the critical concentration points is a semicircular arc as in Fig.6.5(b). For points inside the circular arc two corners are generated, while for points on or outside it one corner only.

Another way to view the phenomenon is to focus on the lengths of the sides of the smaller triangular flap and the associated parts of the square. As you fold down to make the triangular flap the associated parts of the square is divided into two parts, one part that moves and the other part that remains stationary. The length of a leg of the flap, or the part of the side that moved, may be less than half of the side or it may not. If it is, then when the adjoining flap is made part of the stationary side will remain unmoved and become a side of the final polygon (as in Fig.6.7(a)).

Now apply all the above considerations to all the vertices of the square. For now, exclude the case where the point of concentration is the center or a vertex of the square. Select a particular point of concentration; some of the vertices will be nearer, others will be farther from the concentration point. Select a nearest vertex. When this vertex is placed on the concentration point each of the two sides of the square containing this vertex are divided into two parts; one part that moves and the other part stationary. Two cases arise regarding the lengths of the parts that moved.

(i) Both parts are less than half a side. In this case part of both stationary portions will remain exposed, and therefore the resulting polygon will have 4 + 2 or 6 sides and a hexagon is formed.

(ii) Only one part is less than half a side. Part of only one stationary portion will remain unmoved, and therefore the resulting polygon will have 4 + 1 or 5 sides and a pentagon is formed.

In both cases, note that when the opposite vertex is placed on the concentration point both folded-up parts of the sides will be greater than half the side. Hence we can conclude that the type of polygon formed depends on whether or not portions of the sides containing the above two vertices remain unmoved (and therefore become sides of the resulting polygon).

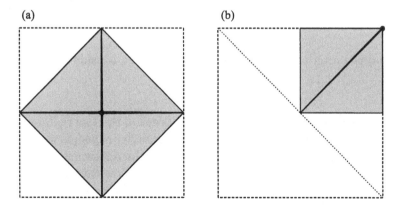

Fig. 6.7   The only positions that quadrilateral would formed.

6.7 What Are the Areas of the Regions?

I noticed from experience that if a point is to be selected at random more people would select a point leading to a hexagon than one leading to a pentagon; and fewer still would select a point leading to a tetragon (or quadrilateral). This leads to the question: are the number of persons so concerned proportional to the areas of the hexagon and pentagon regions?

To answer this question we shall calculate the respective areas.

Since the areas of the petals consist of overlapping areas of semicircles, the areas of the petals may be obtained by deducting the area of the square from the total area of the four semicircles.

Let the length of each side of the square be 1. Then the radius of each semicircle is $\frac{1}{2}$.

The area of each semicircle is $\frac{\pi r^2}{2}$, then $\frac{\pi}{8}$.

Therefore the area of the four semicircles is $\left(\frac{\pi}{8}\right) \cdot 4$, then $\frac{\pi}{2}$.

So the area of the hexagon region, or area of the petals, is $\left(\frac{\pi}{2}\right) - 1$, then about 0.57.

And the area of the pentagon region, or area outside the petals, is $1 - \left(\frac{\pi}{2} - 1\right) = 2 - \frac{\pi}{2}$, roughly 0.43.

It appears that the difference in the areas of the two regions is not significant. So the disparity in the ratio of number of people who choose the respective zones cannot be explained by the difference in areas. A reason to explain the greater attraction of points within the petals, as opposed to outside the petals, might be worth looking into.

Topic 7

HEPTAGON REGIONS EXIST?

Drawing a Distribution Map in the Same Way as Topic 6, but Using Rectangular Pieces of Paper

7.1 Review of the Folding Procedure

Topic 6 introduced the *Vertex-Concentrating Folding* procedure applied to square pieces of paper and determined the polygon types associated with different starting points. In this topic we shall apply the folding procedure to rectangles.

First, we review the procedure.

The basic procedure is as follows. Mark a point on the paper surface. As in topic 6, we shall call this point the *concentration point*. Put one of the vertices of the paper on this point and make a firm crease. Then unfold the paper and place another vertex on the concentration point and make another firm crease. Repeat this procedure to make a crease for all the vertices. In the case of a quadrilateral-shaped paper there should be 4 creases; for a pentagon-shaped paper, 5 creases. These creases and the sides of the paper surround the concentration point to make a polygonal figure. Count the number of vertices and sides in your polygon. What kind of a polygon did you make?

Do this several times, and each time determine the number of sides (that is, determine n) in your polygon.

After you become used to the folding procedure, you can make the creases without unfolding at each step. This way the frame of the polygon stands out more distinctly when the procedure is done. However, two problems may occur with this second procedure (this was also described in topic 6). One problem arises when in the process of placing one vertex on the concentration point, other vertices move too. A second problem arises when in the process of folding successively, part of the original paper may protrude. To overcome these problems, for the first problem, bend the paper to place one vertex on the concentration point without making a firm crease, then position also the moved vertex onto the concentration point, then make firm creases to flatten the figure. For the second problem, tuck the protruding part in. but if this second procedure is confusing, return to the basic procedure.

We study rectangles with given side ratios. Among the popular paper sizes with set ratio of sides is the size A4 copy paper, used in business correspondences. The ratio of the sides of size A4 paper is $1 : \sqrt{2}$.

Take a size A4 sheet of copy paper, select a point and apply the Vertex-concentrating folding. What kind of a polygon did you get? Repeat the folding for other concentration points, and beside each point record the polygon type obtained.

In topic 5 we drew semicircles with centers the midpoints of the sides and diameters the sides to form a four-petal design, and thus divided the square into parts as follows (refer back to Fig.6.6):

- the area inside only one semicircular region (outside the petals).
- the area where two semicircular regions overlap (within the petals).
- points of intersection of two or more semicircular arcs (center, vertices of square).
- all other points on the semicircular arcs.

And we found out that each part induces a particular type of polygon.

Let us divide the rectangular sheet of paper in the same manner. Take a clean sheet of size A4 paper. Draw semicircular arcs, each with center the midpoint of a side and diameter that side, to obtain Fig.7.1. This divides the rectangular region into the following parts:

area a: two areas along the shorter sides inside only one semicircle.
area b: two areas along the longer sides inside only one semicircle.

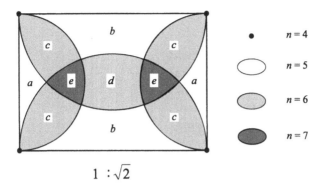

Fig. 7.1 A heptagon region appears a rectangular A4 paper.

area c: four areas emanating from the vertices where two semicircular regions overlap.
area d: one area at the center where two semicircular regions overlap.
area e: two areas near the center where three semicircular regions overlap.
area f: the points on the semicircles.

Note that, unlike the square, there are two different sizes of circles; also an area exists where three circular regions overlap.

I hypothesize that this partitioning of the rectangular region yields a distribution map of the concentration points according to the types of polygons induced.

We now proceed to verify this hypothesis.

7.2   A Heptagon Appears!

In each of the regions a to e mark a few points at random, and with each point as concentration point fold the paper to see what kind of polygon appears. Beside each concentration point record the type of polygon induced.

After obtaining the polygon for some concentration points try to predict the kind of polygon for other points before going through the folding procedure again. You should obtain the following results:

area a: $n = 5$ (pentagon region)
area b: $n = 5$ (pentagon region)
area c: $n = 6$ (hexagon region)
area d: $n = 6$ (hexagon region)
area e: $n = 7$ (heptagon region)

Thus the number of sides n of the polygon induced by a particular concentration point depends on the number of circular regions in which it is located — one, two or three — regardless of the size or position of the circle.

And what about the borders of these areas: what types of polygons do they induce? By selecting concentration points on the borders and carrying out the folding we will find out that points on the border between two adjacent areas induce the same type polygons as the area with smaller n. For example, a pentagon is induced on the border of b ($n = 5$) and c ($n = 6$); also on the intersection point of a ($n = 5$), b and c.

In addition to the above points, we need to check the case when the vertices of the rectangle are concentration points. By folding we see that a quadrilateral is induced; that is, at the vertices $n = 4$.

Note that the center of the rectangle lies in area d ($n = 6$).

7.3 Experimenting with Rectangles with Different Ratios of Sides

Let us now work on rectangles with other ratios of sides.

Figure 7.2(a) shows a rectangle with sides in the ratio 1 : 2. The areas look different from those in Fig.7.1, but their descriptions are the same:

area a: two areas along the shorter sides where no two circular regions overlap.

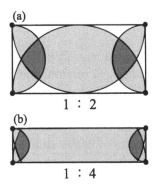

Fig. 7.2 The same pattern is observed even if the ratio of sides of the rectangle is changed.

area b: two areas along the longer sides where no two circular regions overlap.
area c: four areas at the vertices where two circular regions overlap.
area d: one area at the center where two circular regions overlap.
area e: two areas near the center where three circular regions overlap.
area f: the points on the circles.

And the types of polygons induced in the rectangle are the same as those induced in the size A4 paper.

Figure 7.2(b) shows a rectangle that has ratio of sides 1 : 4. Areas (a) and (b) are very narrow or tiny. Nevertheless, the types of polygons induced are still the same.

So for all rectangular regions studied so far, regardless of the ratio of the sides, the parts where no circular regions overlap induce pentagons, the parts where two circular regions overlap induce hexagons, and the parts where three circular regions overlap induce heptagons. When the centralizing point is on a border the type of polygon induced is that of the bordering area with the smallest n.

We found out that in the case of a rectangle four types of polygons are induced: $n = 4$, $n = 5$, $n = 6$, $n = 7$. In the case of a square, three types of polygons are induced: $n = 4$, $n = 5$, $n = 6$. The amazing discovery that for rectangles heptagons are induced motivates us to try other quadrilaterals.

7.4 Try a Rhombus

We explore in the case of a rhombus. The rhombus needs special attention (Fig.7.3). It is not enough to construct circles with centers at the midpoints of the sides and diameters the length of the sides; we must also draw the circumcircle which go through three of four vertices of a rhombus. When a rhombus is not a square, there are four circles which centers are on midpoint of edges and there are two circles which centers are inside of a rhombus.

When the six circles are drawn the rhombus region is divided into 14 parts. In some areas at least two circular regions overlap and at most four circular regions overlap. The two circular regions comprise the quadrilaterals, the three circular regions do the pentagon regions and the four circular regisssssons do the hexagon regions except for the vertices. The vertices which are on the long diagonal line comprise the quadrilaterals and the others do the pentagon regions.

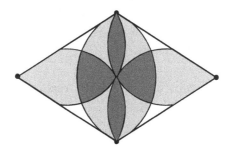

Fig. 7.3 Applying the same procedures to a rhombus.

Topic 8

A WONDER OF ELEVEN STARS

Properties of the Points of Intersection of by Side-to-Line Folds with Reference to an Arbitrary "Mother Line"

8.1 Experimenting with Paper Folding

As you will see later, the title is a very appropriate one. The paper folding is simple but since the subject matter is new some new terms will be brought up.

1. Take a square piece of paper and lay it on the desk. Fold the paper at an arbitrary position as shown in Fig.8.1(a). We shall call the crease formed the **Mother Line** since other folds we make shall be in relation to it. The fact that the Mother Line is arbitrary makes this activity awesome – different experimenters could have different configurations. The Mother Line may intersect adjacent sides of the square, or opposite sides. It may or may not pass through a vertex. It may be short or long, steep or not so steep, closer to the center of the square or farther, and so forth.

In order to distinguish the Mother Line from other creases that we shall make, mark it by some means. One way is to paint the fold with a felt pen as in Fig.8.1(b), then fold it back to make a ridge. A fold made by folding backwards is called a **ridge fold** or **mountain fold**.

Make the crease firm by stroking it with your fingernail.

78   *Properties of the Points of Intersection*

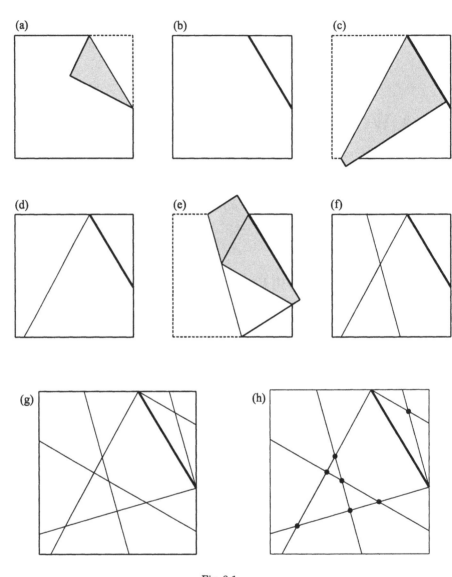

Fig. 8.1

We shall discuss the case where the Mother Line intersects adjacent sides of the square.

2.   With the paper positioned as in Fig.8.1(b), fold the top left side over onto the Mother Line and make a firm crease (Fig.8.1(c)) Thus we have a *side-to-line* fold. Unfold (Fig.8.1(d)). With the left side of the square apply another *side-to-line* fold onto the Mother Line (Fig.8.1(e)). Unfold (Fig.8.1(f)). Do the same with all the other sides or their parts, unfolding each time. Since two sides of the square were divided and two were not, there will be two sides plus four parts of sides. Therefore you should have made six creases (Fig.8.1(g)).

3. Mark the intersections of the side-to-line folds. Mark them with heavy dots (Fig.8.1(h)). In the example, 7 dots were obtained but you may get less. The number of dots will vary from 2 to 7, depending on the position of the Mother Line.

Pause a while and study the arrangement of the dots on your square paper. Can you see some property common to all or most of the dots?

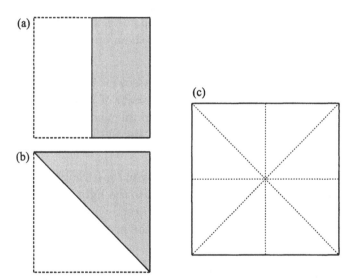

Fig. 8.2

4. Set aside the square paper with the dots. Take another square piece of paper the same size as the first. Make the two diagonal folds and the two book folds making firm creases and unfolding

Properties of the Points of Intersection

after each fold. Thus we have four folds: two book folds perpendicular to each other and two diagonal folds perpendicular to each other. In order to make the folds distinct you may mark them as in step 1, using a different color for each type of fold.

The diagonal and book folds are basic to any origami activity. When one makes a crane or a boat or any other origami model usually the first fold made is a diagonal fold or a book fold. There are two book folds and two diagonal folds. For this reason we shall call these four folds **primary folds**.

And ...

8.2 Discovering

Place the paper with the side-to-line folds and the dots neatly on the paper with the primary creases and hold them up to the light. What do you see?

When this subject was taken up in classes for 9th grade students at the Middle School attached to the University of Tsukuba, Japan, and the students saw that all the points lie on the primary creases, various exclamations of surprise went up: "Ah!" "Really!" "Unbelievable!"

5. If the paper is opaque it is not easy to see through the two pieces of paper, even if they are held up to the light. An alternative is to produce the primary creases directly on the original square as in Fig.8.3. The dots all lie on the primary creases!

A basic principle of the scientific method is that one must avoid making a generalization on the basis of just one activity or experiment. Therefore step 6 is proposed.

6. Take another clean piece of square paper and make another Mother Line with a different position, length or direction. Then apply the sequence of side-to-line folds to obtain the dots on the intersections of the folds, finally construct the primary lines.

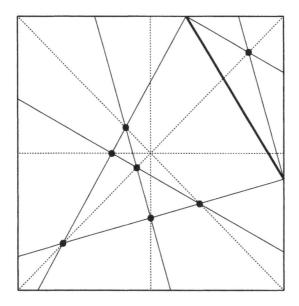

Fig. 8.3 "Oh! There are all intersections on the primary creases!"

Other positions of the Mother Line are shown in Fig.8.4. Repeat the side-to-line folds with other positions of the Mother Line; for example, closer to the edge or on a diagonal. Among the positions of the Mother Line that you tried, which one gave you the most number of intersections?

(a) 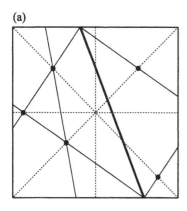   (b)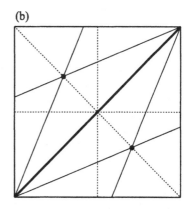

Fig. 8.4

Results of our experiments lead us to believe more strongly the following:

> On a square piece of paper select an arbitrary Mother Line. Construct all edge-to-line creases with reference to this Mother Line. Then all the points of intersection of these edge-to-line creases that appear inside the square lie on the primary creases.

This, however, is still nothing but a hypothesis. Therefore we establish a proof.

8.3 Proof

For the proof we transfer the results of our paper folding to drawings.

In our first example seven points of intersection of the creases were obtained. Let us denote them by P, Q, R, S, T, U, V. Also, let the square be ABCD, and the Mother Line EF; see Fig.8.5. We shall name other points as the need arises.

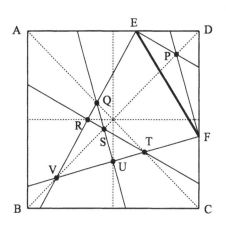

Fig. 8.5 The seven intersection points obtained from Fig.8.1.

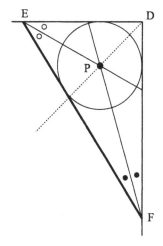

Fig. 8.6 The creases bisect the angles.

Consider first point P, the point of intersection in right △EDF; see Figs.8.5 and 8.6. It is known that the bisectors of the angles of a triangle are concurrent; furthermore, the point of concurrence is the incenter, the center of the circle inscribed in the triangle. Since the two creases intersecting at point P are produced by bisecting ∠E and ∠F of the triangle, then P must lie on the bisector of ∠D as well. But the bisector of ∠D is a diagonal, or a primary crease. Therefore P lies on a primary crease, and P is the incenter of the triangle.

Consider next the point U. Let FE and AB intersect at point G, let FU intersect AB at point H, and look at △GHF (Fig.8.7(a)). Recall that FU was obtained by the edge-to-line fold of FC onto FE. By this folding procedure FU bisects ∠EFC, therefore ∠EFH = ∠HFC. But AB and CD are parallel; therefore the alternate interior angles made by FH are equal, namely ∠HFC = ∠FHG. So ∠EFH = ∠FHG, and △FGH is isosceles with base FH. By the edge-to line fold of AB onto FE, GU bisects ∠G, therefore also the base FH. That is, U is the midpoint of FH. Sides AB, CD and the midline between them are parallel and equidistant, therefore they cut off equal segments on any transversal, in particular on FH. Thus the midpoint U lies on the midline, a primary crease.

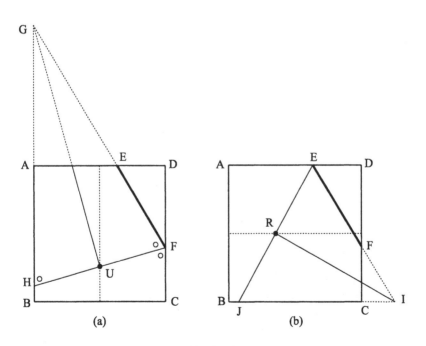

Fig. 8.7

Likewise point R is the midpoint of the base of the isosceles triangle with vertex the intersection of the extensions of the Mother Line EF and the side of the square BC (point I in Fig.8.7(b)), and therefore lies on the midline between sides AD and BC, a primary crease.

The arguments for points Q, T and V are based on another condition of triangles: the bisectors of the exterior angles at two vertices and the bisector of the interior angle at the third vertex are concurrent; the point of concurrence is called an excenter of the triangle. In the case of point Q, looking at right $\triangle$GAE of Fig.8.7(a) together with Fig.8.5, GQ (that is, GU) bisects the interior angle at G and EQ bisects the exterior angle at E. Therefore the intersection, Q, is an excenter and lies on the bisector of the exterior angle at D, which is a diagonal, a primary crease.

Similarly point T is an excenter of the right $\triangle$BFI, and therefore lies on the diagonal through B, a primary crease. And V is an excenter of right $\triangle$EDF so it lies on the bisector of $\angle$D, a diagonal.

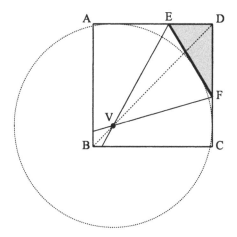

Fig. 8.8   V is an excenter of $\triangle$EDF.

An interesting property of an excenter of a triangle is that it is the center of a circle externally tangent to the associated triangle. That is, V is the center of the circle outside $\triangle$EDF tangent to EF, AD (or AE) and CD (or CF). This turns out to be an extraordinarily large circle, as shown in Fig.8.8. The circle is an excircle of $\triangle$EDF.

When this subject was taken up as an experimental teaching exercise in two high schools, the students had a difficult time since they were not too familiar with excenters and excircles of triangles. Therefore we present another proof here for points V, Q and T that does not make use of these ideas

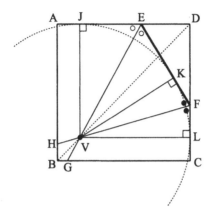

Fig. 8.9 It can be proved without using the "excenter".

Construct the perpendiculars from point V to the side AD, the Mother Line EF and side CD, and let the feet be J, K and L respectively (Fig.8.9). In right △VKF and △VLF, ∠VFK = ∠VFL; therefore the two triangles are congruent and VK = VL. In right △EKV and △EJV, ∠VEK = ∠VEJ; so the two triangles are congruent and VK = VJ. Thus VL = VJ. And since these are perpendicular to the sides of the original square, VJDL is also a square, with diagonal VD coinciding with diagonal BD of the original square. That is, V is on a primary crease.

The arguments for points Q and T are similar.

8.4 More Revelations Regarding the Intersections of the Extensions of the Creases

As mentioned earlier, the number of intersection points that appear on the square varies from 2 to 7 depending on the position of the Mother

Line. If the Mother Line is a little longer than that in Fig.8.3, then the number of intersections that appear becomes 6. (Fig.8.10) All these 6 points undoubtedly lie on the primary creases (Fig.8.11). But why does the number of intersections decrease? The reason is that the two lines that determine point V in Fig.8.5 intersect outside the square. So that if the two lines (creases) are extended the 7th point appears outside the square (Fig.8.12).

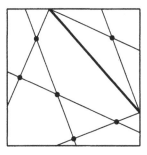

Fig. 8.10 The number of intersections varies according to the position of the Mother Line.

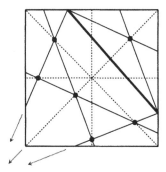

Fig. 8.11 An intersection seems to appear when the creases on the left are extended.

That an intersection point falls outside of the square motivates us to extend the other lines (creases) outside the square and study their intersections. In Fig.8.12 four new intersection points appear, W, X, Y and Z. But, contrary to expectation, none of them falls on any extension of a primary crease!

However, I had a feeling that there might be some connection with

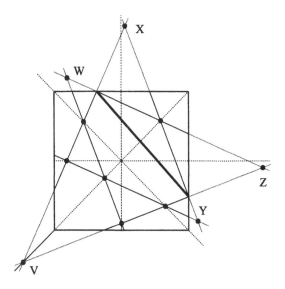

Fig. 8.12 The intersections increase further if the creases are extended.

the primary creases. I played around with another square piece of paper with primary creases, placing it in different positions around the original working square. Eventually I found that if I set the moving square side by side with the original square on the right, two points fall on the diagonals. Excitedly I continued moving the square and found that if it is placed above the original square the two other points fall on the diagonals of the moving square (Fig.8.13). I was so thrilled to see these results, and reexamined the phenomenon on other square pieces of paper with different Mother Lines. It turned out that all the intersections of the extended creases fell on the diagonals.

> On a square piece of paper construct the edge-to-line creases with reference to an arbitrary Mother Line. If the working square and other squares of the same size are laid together in a tiling pattern, the points of intersection of the edge-to-line folds fall on primary creases of the squares.

You may wish to explore further and look for other relations, using other positions of the Mother Line. It is hoped that these manipulative activities will arouse your curiosity and motivate you to continue looking for other relations between the intersections of edge-to-line folds on

88   Properties of the Points of Intersection

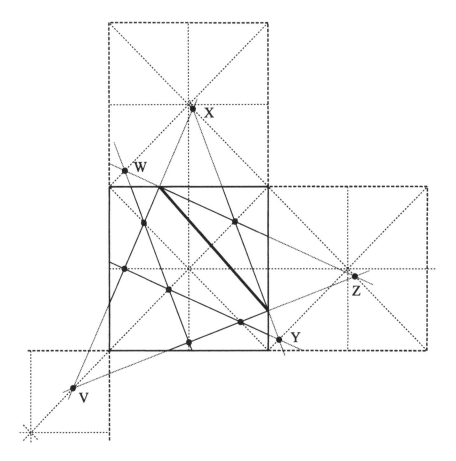

Fig. 8.13   The intersections fall on the diagonals of the moving squares.

a square piece of paper and the primary lines.

Again, remember that a collection of several experiments does not constitute a proof. But solving a puzzle can be so much fun, so I do not always give the proofs of results obtained in class. Giving the proof would be removing the suspense, like telling who the criminal is in a detective story before knowing the story. However, as this is an explanation book for teachers, I should not omit the proofs. Here then is a proof for the last observation above.

8.5 Proof of the Observation on the Intersection Points of Extended Edge-to-Line Creases

In this demonstration we make use of incenters and excenters of triangles Consider points W and Y (Fig.8.14). Extend the Mother Line to intersect the extensions of the sides BA and BC of the square.

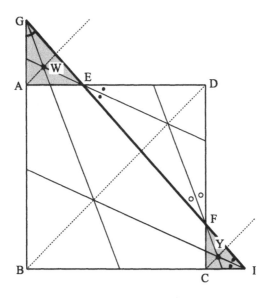

Fig. 8.14 W and Y are the incenters.

Let the points of intersection be G and I respectively. Look at right △GAE. We pointed out that GW bisects ∠G. Also EW bisects ∠AEG (or ∠DEF). Therefore W is the incenter of triangle GA, and W should lie on the bisector of right ∠A, which is a diagonal of the upper square, a primary crease.

In the same way Y is an incenter of right △FCI, and therefore lies on a diagonal of the moved square alongside.

Next consider points X and Z with respect to right △EDF (Fig.8.15). Regarding point X, FX bisects interior angle F, EX bisects the exterior angle at E. Therefore X is an excenter of △EDF and lies on the bisector of the external angle at D, namely a diagonal of the moved square, a

90 Properties of the Points of Intersection

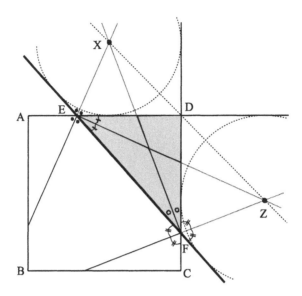

Fig. 8.15   X and Z are the excenters.

primary crease.

Likewise, Z lies on the bisector of the interior angle at E and on the bisector of the exterior angle at F. So Z is an excenter of $\triangle EDF$ and lies on the bisector of the exterior angle at D, a diagonal and therefore a primary crease of the moved square.

If the topic of excenters has not been taken up by the class, the proposition regarding points X and Z can be proven using triangle congruence in the same manner as in Fig.8.9.

We close this topic with a question. Since there are 6 side-to-line creases, then there should be 15 pairs of creases (that is, 6 creases taken 2 at a time)[4]. We discussed 11 pairs of creases (that is, 11 intersections). Where are the other 4?

---

[4] $_6C_2 = 15$

8.6 The Joy of Discovering and the Excitement of Further Searching

Several years ago, while doing origami aimlessly, the phenomenon described in this chapter of side-to-line folds with reference to an arbitrary line came to my mind. And it gave me great delight to study different situations, folding with many pieces of paper while changing the position of the Mother Line. I really felt as if this joy would have faded if the mathematical principles were pointed out too early. So, in a class, to let students feel similar joy and excitement. I give them time just to folding the paper for a while,

When this lesson is used with a class of students, using transparencies with an overhead projector could be useful. Positioning and overlapping the transparencies could enhance visualization, and therefore may heighten student interest. Some examples of transparencies are: (a) a drawing of a square with the creases as in Fig.8.1(g); (b) a drawing with the marked intersection points; (c) one or more drawings with the creases extended; (d) a drawing with the intersection points on the extended lines.

Topic 9

WHERE TO GO AND WHOM TO MEET

Distribution of the Position of a Generating Vertex According to the Different Groups of Vertices That Can Meet It by a Single Fold

9.1 An Origamics Activity as a Game

The title of this topic may sound perplexing and unclear, but once you know the rules of the activity you will realize that it is very appropriate.

This topic is similar to topic 6 in that the investigation of relationships between certain points leads to the emergence of patterns. And while the topic does not intend to compete with other origami activities in creating works of art, the art it does create has its own merits.

9.2 A Scenario: A Princess and Three Knights?

Place a sheet of square paper on the table with the colored or printed side underneath. First we give names to the four vertices. The idea is to give a personality to each of the vertices in order to make the game interesting. One vertex may be called Princess and the other three vertices Knights. When the topic first came to my mind I called the game "the HG game", where one vertex was Host and the other three were Guests. On another occasion, when I was invited to give a talk to a group of 7th grade (first year middle) students and their teachers, I named one corner Self and the other three corners Friends. Other names may be used.

94   Distribution of the Position of a Generating Vertex

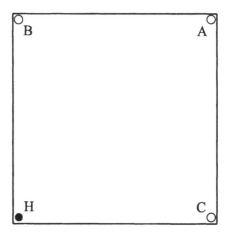

Fig. 9.1   Label the four vertices. H represents the "Host". A, B and C represent the "Guests".

9.3   The Rule: One Guest at a Time

Let us use the names Host and three Guests (Fig.9.1). When a fold is made so that the host (H) is moved, part of the colored (or printed) underside of the paper emerges - let us call the colored part that emerges the domain of the host. In Fig.9.2 the domain of the host is triangular in shape. Likewise, as each guest is moved its domain appears.

The host may meet with a guest if, when the guest approaches the host, its domain does not invade that of the host. The aim of the game is to determine whether, for a particular position of the host, any of the guests may approach the host one by one.

Figure 9.2(b) shows that the host can meet with guest A. Figure 9.2(c) shows that the host can meet with guest B. But in Fig.9.2(d) guest C has gone as close as possible without invading the domain of the host; while in Fig.9.2(e) if guest C forces himself on the host the domain of the host is invaded. Thus the host cannot meet with guest C. Figs.9.2(b) and 9.2(c) are marked with "○" to indicate that guests A or B can meet with the host, while Figs.9.2(d) and 9.2(e) are marked with " × " to indicate that guest C cannot meet with the host. The outcomes may also be indicated symbolically as A(○), B(○), C(×), and the position of H as type [AB].

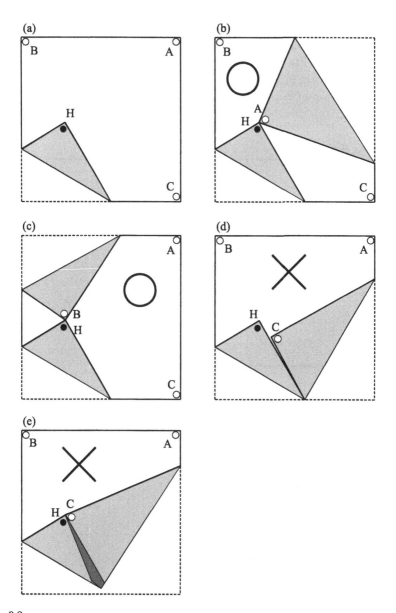

Fig. 9.2
(a) The host moves to a point.
(b) Guest A can meet the host without invading the domain of the host. Mark with "○".
(c) Mark with "○".
(d) Guest C cannot meet with the host. Mark with "×".
(e) If guest C force to meet the host, it is necessary to bend the domain of the host. Mark "×".

Figure 9.3 shows the host in another position. In this case guest B cannot move because it is in the domain of the host (Fig.9.3(a)). Guest A cannot reach H (Fig.9.3(b)). But guest C can meet the host H without any difficulty (Fig.9.3(c)). Hence the outcomes are A(×), B(×), C(○); and the position of H is type [C].

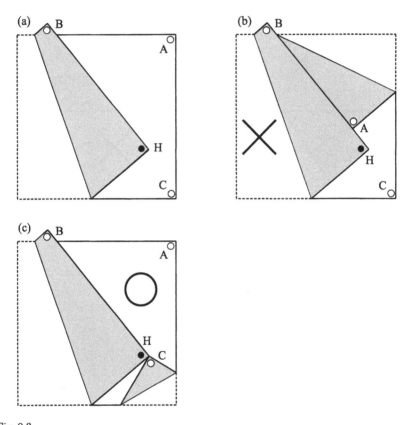

Fig. 9.3
(a) Guest B cannot move, hence ×.
(b) Guest A cannot reach H, hence ×.
(c) Guest C can meet with H, hence ○.

Practice folding the paper a few more times, placing the host H in different positions and determining who of the guests can meet H. Label each position of H according to type.

9.4 Cases Where no Interview is Possible

We have seen that whether or not a Host may meet with a particular guest depends upon its position. For each guest there are two cases: "able to meet" or "not able to meet". So theoretically, for three guests there are $2 \times 2 \times 2$ or 8 cases. These are indicated in the following table.

Table 1

Number	Type	A	B	C	In Fig.9.4
1	[ABC]	○	○	○	●
2	[AB]	○	○	×	○
3	[AC]	○	×	×	▲
4	[BC]	×	○	○	Except in Fig.9.4
5	[A]	○	×	×	■
6	[B]	○	○	×	△
7	[C]	×	×	○	□
8	0	×	×	×	×

You will find out that, while theoretically there are 8 types, in actual practice not all the types can occur. There exists no position of H of type [BC] (or number 4).

9.5 Mapping the Neighborhood

If you have not done so yet, fold the paper a few more times, placing the host H in different positions and determining who of the guests can meet H. Label each position of H according to type, such as [AB] or [C] for the positions of the host in Figs.9.2 and 9.3 respectively, or by numbers, 2 and 7 respectively; see Fig.9.4. If your paper becomes too crumpled use a new piece.

You might think of the piece of paper as a village consisting of addresses 1 to 8, except 4, or addresses [ABC], [AB], ..., [0], except [BC].

As a method of investigation, this repeated folding activity may be done in several ways. One way would be to fold at random and mark the position of H each time. But as an investigation it would be wise to

98   Distribution of the Position of a Generating Vertex

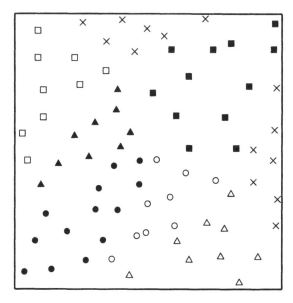

Fig. 9.4   While changing the position of the host at random, check whether the host can meet with A, B and C. Label each position of H according to the type listed in the Table.

plan a system for the positioning of H. For instance, first set positions of H on the diagonal lines, on the midlines, or on the edges. Or construct a grid on the piece of paper and set positions of H on the grid points. What is required most is patience and perseverance.

As you progress in the paper folding, study how the map of the village builds up. Use your powers of analysis and imagination to dig up emerging shapes and patterns. Pay particular attention to the boundaries.

After you have done much of the mapping (perhaps by then the pieces of paper will be in a miserably crumpled state), let us transfer the outcomes (i.e., addresses) to another piece of paper. Using pens of different colors for color coding of different address types would make the map more attractive and patterns more vivid.

Figure 9.5 shows the map of the distribution of H according to address types. Here the original position of H is at the lower left vertex. The figure is an unexpectedly pretty pattern. It consists of two semicircles with radii one half of the side or the square paper centered at

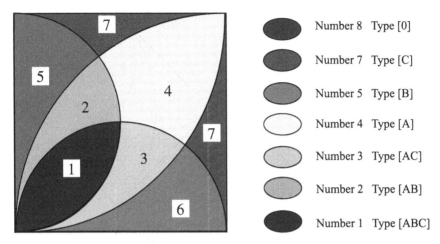

Fig. 9.5 Classify the outcomes, transfer it to another piece of paper, and examine whether the figure is correct. This pretty pattern can be obtained.

the midpoints of the sides and through the original position of the host, and two quarter-circles with radii the side of the paper centered at the vertices adjacent to the host's original position. So the four arcs all contain the point H. Addresses [ABC] are in the region where the four circles overlap - the happy address where the host can meet any one of the three guests.

9.6   A Flower Pattern or an Insect Pattern

Rotate the paper so that the original position of the host is down and the corresponding diagonal is vertical; do you see the flower or leaf design? Now turn the paper upside down. Do you see the cicada or fly? Its wings appear half open.

9.7   A Different Rule: Group Meetings

For variety a different scenario is used here. Bringing up other scenarios enables the teacher to select what he/she thinks is more suitable for his/her learning group.

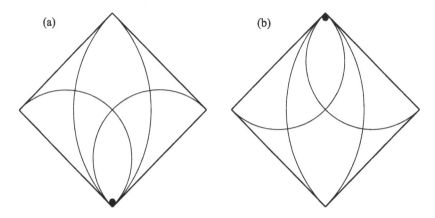

Fig. 9.6 (a) Do you see the flower or leaf design? (b) Do you see the cicada or fly?

I sometimes deal with animal behavior in my lectures in biology. Whenever I explain the lesson in terms of ♂ (male) and ♀ (female) relationships an air of attentive quiet prevails in the class. Figure 9.7 shows a situation involving one ♀ (female) and three ♂ (males). Here ♀ has moved from her original position although in the dotted area and therefore has created a domain for herself. And each ♂ is waiting at the three corners P, Q and R, watching for a chance to approach ♀. The area enclosed by the dotted boundary is that obtained in the first game (Fig.9.6), where any ♂ can meet ♀. In Fig.9.7(b) at once ♂ P makes contact, and he stays there with no intention of moving away. In this state of things neither ♂ Q nor ♂ R can meet ♀ because their actions are blocked by the domain of ♂ P. Thus ♂ P has exclusive rights to ♀.

A different situation is shown in Fig.9.7(c), where ♀ remains in the same position, but if ♂ R makes contact more quickly than ♂ P it blocks ♂ P. But ♂ Q can still meet ♀, as seen in Fig.9.7(d). Thus a triangular affair happens among ♀, ♂ Q and ♂ R.

Figure 9.8 shows another situation where ♀ has moved to a different position. Here ♂ Q cannot meet ♀. But both ♂ P and ♂ R can contact ♀, no matter who of the two made contact first; neither is able to block the other.

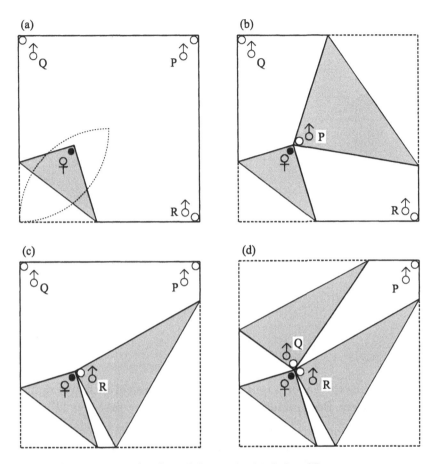

Fig. 9.7 There are one ♀ (female) and three ♂ (male), P, Q and R.
(a) ♀ moves into the dotted area. Each ♂ looks for a chance to approach ♀.
(b) Immediately ♂ P makes contact and he stays there with no intention of moving away. Neither Q nor R can meet ♂ because their actions are blocked by the domain of ♂ P.
(c) ♂ R can make contact more quickly than ♂ P.
(d) ♂ Q can still meet ♀, so a triangular affair happens among ♀, ♂ P and ♂ Q.

9.8 Are There Areas Where a Particular Male can have Exclusive Meetings with the Female?

In order to investigate this question do more experimenting, folding a square piece of paper and unfolding repeatedly for different positions of

102   *Distribution of the Position of a Generating Vertex*

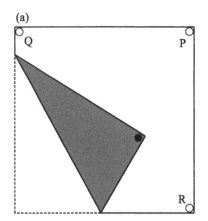
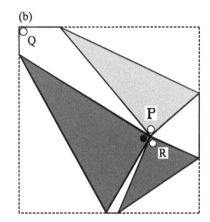

Fig. 9.8
(a) If ♀ moves to this position, both ♂ P and ♂ R can contact ♀, but ...
(b) even if one of them makes contact first, he is not able to block the other.

♀. By the time the paper is quite crumpled the area of exclusive right of each ♂ appears. Be reminded about the "first-come-first-served" rule.

Spend time on paper folding, do not rush to conclusions. The joy of discovery may be felt more strongly when a conclusion is obtained as a result of patiently repeated experiments rather than through logical reasoning from the start.

Figure 9.9(a) shows the area where ♂ P is first-comer and has exclusive meetings with ♀, while Fig.9.9(b) and (c) show the equivalent areas for ♂ Q and ♂ R respectively. Observe that the respective areas for ♂ Q and ♂ R are symmetric images with axis of symmetry a diagonal of the square. Each area is bounded by arcs of two circles with respective radii $\frac{1}{2}$ and 1. Observe also that these areas are independent of those on the map of Fig.9.5.

The overlap of the three areas, shown in Fig.9.9(d), is the very field of battle where the three ♂ 's must fight desperately to reach ♀ first. Any ♂ can bar the other two by the rule of first-come-first-served.

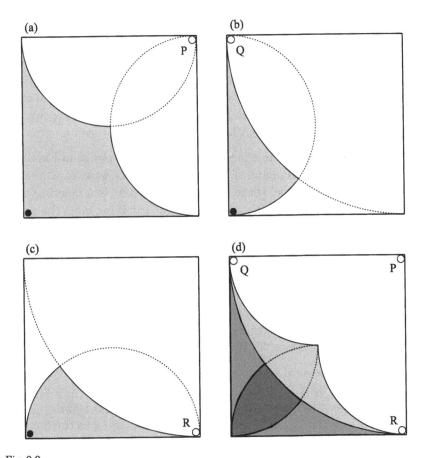

Fig. 9.9
(a) The area where ♂ P is first-comer and has exclusive meetings with ♀. The dotted line is an auxiliary line which explains the evolution of the area.
(b) The area where ♂ Q is first-comer and has exclusive meetings with ♀.
(c) The area where ♂ R is first-comer and has exclusive meetings with ♀.
(d) The overlap of the three areas is the very field of battle.

9.9 More Meetings through a "Hidden Door"

While continually playing with the paper, all of a sudden I realized that there is also the case where ♀ would move to points outside the paper; this situation is shown in Fig.9.10(a). With ♀ in this position, ♂ R cannot move at all since it has already been taken into the domain of ♀. And ♂ P, although having a little freedom to move around, can never reach ♀. Also ♂ Q, although having still more freedom, is blocked by the domain of ♀.

But if the paper is folded away from you to make a mountain fold, ♂ Q could meet ♀ (Fig.9.10(b)). One might explain that ♂ Q enters through a "hidden door" to meet ♀. Now, what is the extent of the area wherein ♂ Q can meet ♀ in the above way? Again we can repeatedly fold and unfold the paper for different positions of ♀ to seek an answer.

Make a fold to move ♀ to a location outside the paper as in Fig.9.10. The flap made is the domain of ♀. Then move the nearest ♂ vertex through the "hidden door" (that is, fold backwards). Is a meeting possible? If yes, label the location as "Successful" (S); if no, label the point as "Unsuccessful" (U).

To make a record of the result, transfer the setup to a drawing. On another sheet of paper draw a square of the same size as the square paper, mark the position of ♀, and label this position as "S" or "U".

Repeat this paper folding activity several times for other locations of ♀ and for possible meetings with ♂ Q. Mark the positions of ♀ on the drawing and label each as "S" or "U".

After folding several times it will come to light that the condition for a location to be a favorable meeting place is that the two creases made should not intersect in the paper region (see Fig.9.10(b)).
[Regard the location of ♀ as head and the opposite vertex on the domain - that is, the far end of the crease - as the *foot*. After folding several times it will come to light that the condition for a location to be a favorable meeting place is that the two creases made should not intersect before the foot (see Fig.9.10(b)).]

The investigation can be continued in like manner for possible meetings between ♀ and ♂ R.

Studying the distribution of S's and U's on your drawing, you will see that the bounds of the S points are the two narrow shaded areas shown in Fig.9.10(c). Each circular segment is bounded by a circular arc with radius the distance between a vertex on this circular arc and the midpoint of the opposite side, and center the aforementioned midpoint. The two circular segments correspond to meetings with ♂ Q and with ♂ R respectively.

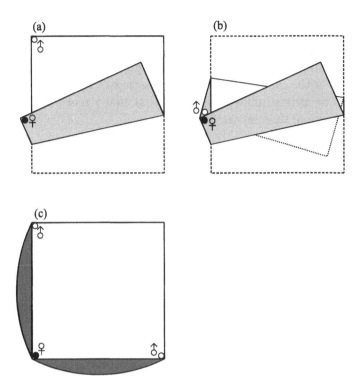

Fig. 9.10

(a) In the case where ♀ moves to points outside the paper; In this position, only the nearest ♂ vertex can move through a "hidden door".
(b) This ♂ can meet with ♀ from the hidden side of the paper.
(c) The area which can be met through "hidden door" is crescent - shaped. The center of the circle is a midpoint of the opposite side.
(From "Oru" vol.11)

So if the female moves into this narrow shaded area the male ♂ which happens to be closest will be able to meet the female through the "hidden door", but she will be out of reach of the other males. Thus there is a small private hideaway for the pair.

I injected a human dimension into the discussion to give some personality to the discussion and to arouse interest. Our hearts are warmed to the situation of the male and female couple (in contrast to a situation between two points). It is difficult to feel poetic and sentimental over a discussion in mathematics or in animal science when it is so technical.

I was tempted to continue the discussion to other situations - for instance, change the rules, or use a rectangular playing field instead of a square. For the case of the rectangle, the task is a little more difficult, but I was surprised anew to find out that many more mathematical principles reveal themselves.

Clearly origami can be a very fertile ground for bringing out and applying many mathematical relationships.

Topic 10

INSPIRARATION OF RECTANGULAR PAPER

Dividing a Rectangular Sheet of Paper into a Number of Equal Parts Horizontally and Vertically

10.1 A Scenario: The Stern King of Origami Land

The whole country of Origami Land is dedicated to the appreciation and creation of origami art with square paper, and the king enforced this strictly as a law. One day a wanderer was caught working with rectangular paper, and he was arrested. When brought before the king he was scolded, "You cannot even fold origami paper properly! You are just amusing yourself by folding aimlessly, supposedly in the interest of mathematics. It is outrageous the way you desecrate origami! " Then he turned to the soldiers, "Throw him in jail and give him nothing to eat!"

But then in a change of heart he turned back to the wanderer and said, "I shall give you a task; if you accomplish it you will be fed and be freed. This is the task:

"You must find a way to divide your rectangular sheet of paper into 17 equal parts horizontally and 17 equal parts vertically."

And so the wanderer was thrown into jail with only the one sheet of paper – not even a pencil. In despair he resigned himself to starving and death. But after a while he overcame this feeling; he had a sharp mind and an idea flashed.

108   *Dividing a Rectangular Sheet of Paper*

Let us follow a line of reasoning.

10.2 Begin with a Simpler Problem: How to Divide the Rectangle Horizontally and Vertically into 3 Equal Parts

In this chapter we shall use international standard-sized rectangular paper; that is, the ratio of length to width is $\sqrt{2} : 1$. A popular size is A4 paper.

We want to obtain a node which divides the paper horizontally and vertically into 3 equal parts[5]. Let us call each intersection point a node. One can see that if the position of one node is known then a folding procedure can be devised. So as a start we need to locate a node.

Fig.10.1 shows three ways to locate one node. These methods were obtained from discussions and readings. Method (a) was obtained in Topic 3; method (b) was verified in the magazine Origami Tanteidan (in Japanese), No. 79 (2003)[6], and method (c) shall be explained below. The rectangles were drawn so that the positions of the three nodes are aligned.

We now explain why the procedure of Fig.10.1(c) works; the explanations of the procedures of figures (a) and (b) are left to the reader.

For our future discussions we rotate the rectangle of Fig.10.1(c) so that the midpoint is on the upper side. And we label the points.

The position of the intersection point can be described by using a coordinate system. Set the axes as in Fig.10.3, with two sides of the rectangle on the axes and their common vertex the origin.

Be reminded that, since the paper we are using is an international standard size, the ratio of the long side to the short side of the rectangle is $\sqrt{2} : 1$.

---

[5]See Fig.10.4.
[6]In this book, it is not translated.

INSPIRARATION OF RECTANGULAR PAPER 109

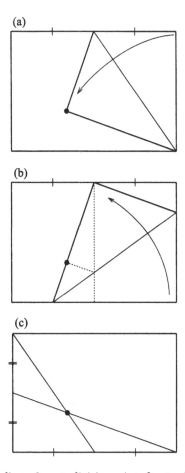

Fig. 10.1 Methods of finding a 3-parts division point of a standard rectangle.
(a) Position of the upper right vertex, as the paper is folded down along the line joining the midpoint of the upper side and the lower right vertex.
(b) Shift-position of the midpoint of the lower side when paper is folded so that the lower right vertex falls on the midpoint of the upper side.
(c) An intersection point is obtained when two lines joining the midpoints of two adjacent sides with a vertex on the opposite side respectively, meet.

We obtain the equations of the lines.

Line ED: $$y = \frac{1 - \frac{1}{2}}{\sqrt{2} - 0}(x - 0) + \frac{1}{2} \quad \text{or} \quad y = \frac{x + \sqrt{2}}{2\sqrt{2}} \quad (1)$$

Line FB: $$y = \frac{2x}{\sqrt{2}} \quad \text{or} \quad y = x \quad (2)$$

110   *Dividing a Rectangular Sheet of Paper*

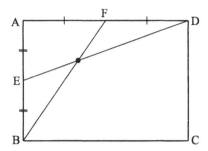

Fig. 10.2   The rectangle of Fig.10.1(c) was rotated so that the midpoint of the longer side is on the upper side.

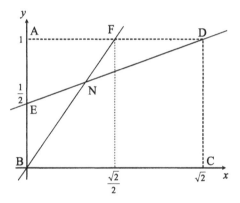

Fig. 10.3   Assuming the width is 1 and length is $\sqrt{2}$, the coordinates of the points are as shown in the figure.

Setting (1) = (2), we obtain the coordinates of the intersection N.

$$x = \frac{\sqrt{2}}{3}; \quad \text{therefore} \quad y = \frac{2}{3}.$$

These coordinates show that x is $\frac{1}{3}$ of length AD, and y is $\frac{2}{3}$ of length AB, and the ratio of the parts into which the paper is divided is 1 : 2 horizontally from left to right and 2 : 1 vertically upwards.

Therefore they suggest the following folding procedure for dividing the rectangle.

For the sake of brevity we shall call the point obtained in Fig.10.4

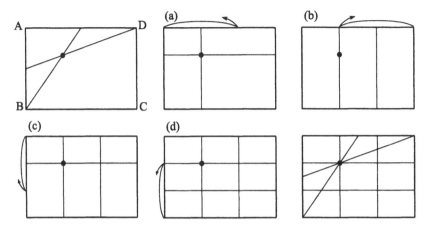

Fig. 10.4 Based on the position of the point of intersection in Fig.10.3, the rectangle could be divided into 3 equal parts horizontally and vertically.

(1) Make a fold through N parallel to CD and unfold.
(2) Align CD on N and make a fold parallel to AB; C and D should lie on the sides of the rectangle. Unfold.

Steps (1) and (2) divide the rectangle horizontally into 3 equal parts.

(3) Make a fold through N parallel to AD and unfold.
(4) Align BC on N and make a fold parallel to AD; B and C should lie on the sides of the rectangle. Unfold.

Steps (3) and (4) divide the rectangle vertically into 3 equal parts.

above as a "3-parts division point" for dividing the rectangle into 3 equal parts horizontally and vertically.

10.3 A 5-parts Division Point; the Pendulum Idea Helps

It is interesting to note that a 5-parts division point may be obtained by a slight change in the procedure for the 3-parts division point. Construct the line ED as in the method for a 3-parts division.

Imagine swinging the line FB to the right, as a pendulum suspended from F. Point B will fall on point C. The intersection of ED and FC is a 5-parts division point; that is, it can be used to divide the paper into 5 equal parts horizontally and 5 equal parts vertically.

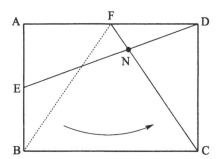

Fig. 10.5   The line FB is "swung" to the right in pendulum fashion to the position FC.

Before verifying this procedure, examine the position of the 5-part division point in Fig.10.5. How many parts (out of 5) do you think will lie to the left of the division point? To the right? Above? Below?

To verify the procedure of Fig.10.5 we use a coordinate system as in the previous section.

Line ED: $$y = \frac{x + \sqrt{2}}{2\sqrt{2}} \qquad (1)$$

Line FC: $$\frac{y}{x - \sqrt{2}} = \frac{1}{\frac{\sqrt{2}}{2} - \sqrt{2}} \quad \text{or} \quad y = -\sqrt{2}x + 2 \qquad (2)$$

(see Fig.10.6).

Solving these simultaneous equations to obtain the coordinates of the point of intersection, we arrive at

$$x = \frac{3\sqrt{2}}{5}, \quad y = \frac{4}{5}.$$

That is, x is $\frac{3}{5}$ of the length BC and y is $\frac{4}{5}$ of the length AB (see Fig.10.7).

This means that by using this intersection point the paper can be divided in the ratio 3 : 2 in the horizontal direction (left to right) and 4 : 1 in the vertical direction (upward).

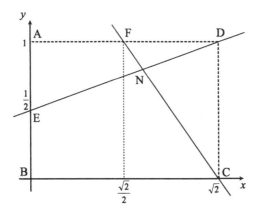

Fig. 10.6 Assuming the width is 1 and length is $\sqrt{2}$, the coordinates of the points are as shown in the figure.

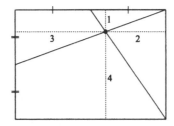

Fig. 10.7 He grid lines through the point of intersection divide the rectangle horizontally in the ratio 3 : 2 and vertically in the ratio 4 : 1, based on the point of intersection in Fig.10.5.

These relations suggest a procedure for dividing the rectangle; see Fig.10.8.

Let us study the folding procedure for 5-parts division more carefully. Recall that a rectangle can be divided equally into $2, 4, 8, \ldots, 2k$ by folding in half, then folding in half again, then folding in half again, and so forth. In Fig.10.8, note that the rectangle would have 1 part above the intersection; this part is obtained in step (1). Note also that there should be 4 equal parts below the intersection; these 4 equal parts are obtained in steps (2) and (3) using the "fold-and-fold-again" method. For the vertical division Fig.10.8 shows that there should be two parts on the right; these are obtained in step (4). Finally there

Dividing a Rectangular Sheet of Paper

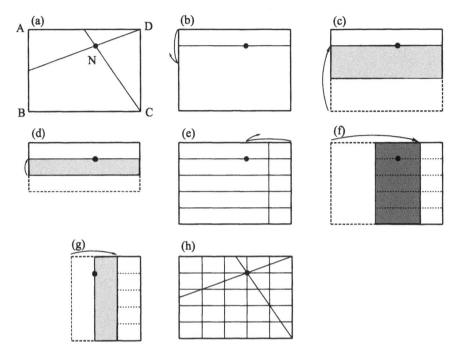

Fig. 10.8   Based on the point of intersection in Fig.10.7, the rectangle is divided into 5 equal parts horizontally and vertically.

(1) Make a book fold through the division point N parallel to the side BC and unfold.
(2) Align BC on N and make a fold parallel to AD. Do not unfold.
(3) Align the new fold in step (2) with N and make another fold parallel to AD. Unfold all.

Steps (1) to (3) divide the rectangle vertically into 5 equal parts.

(4) Align CD on N and make a fold parallel to AB. Unfold.
(5) Align AB on the fold of step (4) and make a fold parallel to CD. Do not unfold.
(6) Align the new fold in step (5) and again fold in half. Then unfold all.

Steps (4) to (6) divide the rectangle horizontally into 5 equal parts.

should be 3 equal parts to the left of the intersection. We take one part to the right of the intersection which was marked with a crease in step (4) so that there would be 4 equal parts to the left of the new crease. So this part to the left of the crease can be divided into 4 equal parts by the "fold-and-fold-again" method, as in steps (5) and (6).

Whenever the term of a ratio is odd, this maneuver of attaching parts on the other side of the intersection to make $2k$ folds can be used.

10.4 A Method for Finding a 7-parts Division Point

Having arrived at a method for 3-parts and 5-parts division, the author felt elated. But then for the 7-parts problem for a while he felt at a loss. But then an idea flashed: use other points on side BC and try a procedure similar to the 3-parts division procedure.

We try a point on side BC whose distance from the origin is $\frac{1}{4}$ of the length of the long side; call this point G (Fig.10.10).

We can find the $\frac{1}{4}$ position of the bottom side by making a book fold through the midpoint to divide the rectangle into 2 equal parts, then making another book fold parallel to the first to divide the rectangle into 4 equal parts. Each of the points on the side made by the folds is called a *quadrisection* point of the side.

Another method for finding a quadrisection point for international standard paper is shown in Fig.10.9. When the paper is folded so that opposite vertices coincide, the fold made marks precisely a quadrisection point on each of the two long sides; see Fig.10.10. The proof of this procedure is left to the reader.

Going back to our main discussion, in Fig.10.10 we look for the position of the new point of intersection.

Again we use a coordinate system

Line ED: $\qquad y = \dfrac{x + \sqrt{2}}{2\sqrt{2}}$ (1)

Line FG: $\qquad \dfrac{y}{x - \dfrac{\sqrt{2}}{4}} = \dfrac{1}{\dfrac{\sqrt{2}}{2} - \dfrac{\sqrt{2}}{4}}\quad$ or $\quad y = (2\sqrt{2})x - 1.$ (2)

Solving these equations simultaneously to obtain the coordinates of the point of intersection N,

$$x = \frac{3\sqrt{2}}{7} \quad \text{and} \quad y = \frac{5}{7}$$

116   *Dividing a Rectangular Sheet of Paper*

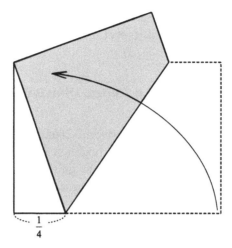

Fig. 10.9   The method of finding the quadrisection point near the lower left vertex of the length of a standard rectangle.

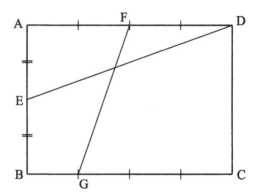

Fig. 10.10   The original line ED is maintained. Join FG and this would give us a new point of intersection.

(see Fig.10.11).

That is, x is $\dfrac{3}{7}$ of length BC and y is $\dfrac{5}{7}$ of length AB. This means that by using the grid lines through this intersection point the paper is divided in the ratio 3 : 4 in the horizontal direction and 5 : 2 in the vertical direction.

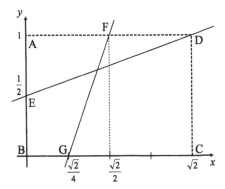

Fig. 10.11 Assuming the width is 1 and length is $\sqrt{2}$, the coordinates of the points are as shown in the figure.

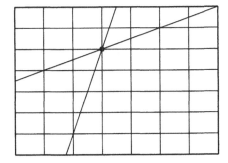

Fig. 10.12 Based on the position of the node in Fig.10.10, the rectangle is divided into 7 equal parts horizontally and vertically by the creases.

The description of the folding procedure based on Fig.10.12 is left to the reader.

10.5 The Investigation Continues: Try the Pendulum Idea on the 7-parts Division Method

Up to this point it has been possible to find the division points for 3-parts division, 5-parts division, 7-parts division and 9-parts division by similar methods. The methods for 3-parts division and 5-parts division form a mutual pair where a division point in one pair is obtained by "swinging" one of the lines as in a pendulum. Pursuing this line of thought, the 9-parts division point might be obtained by "swinging" the line of the 7-parts division procedure (see Fig.10.13); let us find out.

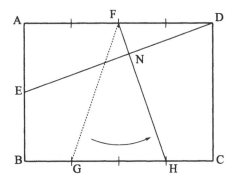

Fig. 10.13   The line FG is swung like a pendulum to line FH.

This new line intersects the side of the rectangle at the point H, a new quadrisection point such that the length CH is $\frac{1}{4}$ of the length of BC, or BH is $\frac{3}{4}$ of the length of BC.

A folding procedure to find H uses the "fold-and-fold again" idea and was described earlier. Another procedure for standard paper is one similar to Fig.10.9. See Fig.10.14.

Figure 10.13 shows how our rectangle would appear if we fold along ED and FH. Pause a while and study the figure closely, making tentative measurements with the fingers. Make a hypothesis: from the appearance, into what ratio do you think the intersection point divides the rectangle horizontally? How about vertically?

We can verify our hypothesis by calculations (see Fig.10.15),

$$\text{Line ED:} \qquad y = \frac{x + \sqrt{2}}{2\sqrt{2}} \qquad (1)$$

$$\text{Line FH:} \qquad y = -\frac{4x}{\sqrt{2}} + 3. \qquad (2)$$

In other words, x is $\frac{5}{9}$ of BC, and y is $\frac{7}{9}$ of AB.

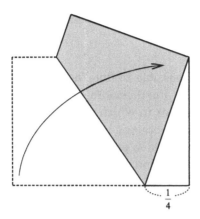

Fig. 10.14 The method of finding the quadrisection point near the lower right vertex of the length of a standard rectangle.

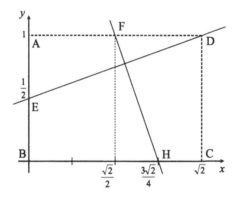

Fig. 10.15 Maintain the original line ED,. Join FH and thus a new intersection point is obtained.

These calculations show that the grid lines through the intersection point divide the rectangle into 9 parts, horizontally in the ratio 5 : 4 and vertically in the ratio 7 : 2.

Does this agree with the hypothesis we made earlier?

The folding procedure is left to the reader.

120   Dividing a Rectangular Sheet of Paper

10.6 The Search for 11-parts and 13-parts Division Points

Up to this point it is possible to find the 3-parts, 5-parts, 7-parts and 9-parts division points. Furthermore, the 3-parts and 5-parts division points form a mutual pair in which the procedure for obtaining the 5-parts division point was obtained by using a pendulum idea on the procedure for the 3-parts division point; likewise, also the 7-parts division point and the 9-parts division point.

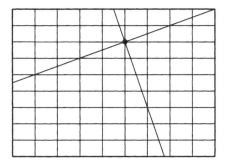

Fig. 10.16   Based on the position of the node in Fig.10.13, the rectangle is divided into 9 equal parts horizontally and vertically by the creases.

If we pursue this argument, we might think: might the 11-parts and 13-parts division points also be a mutual pair? With line ED fixed, what other line intersecting line ED should be taken?

One way to look for the second line would be to work backwards: (1) start from the desired grid (this can be done using geometry software), (2) draw the first line, the line from the midpoint of the short side to the upper opposite vertex (line ED), then (3) look for a second line from the top midpoint that would intersect the first line at a node of the grid and would intersect the bottom side of the rectangle at a convenient point - this point is what we are looking for.

If you do this carefully it appears that the distance from the left vertex of the point on the bottom side might be $\frac{1}{3}$ of the length of the long side (a trisection point). And its pair obtained pendulum fashion would be the other trisection point $\frac{1}{3}$ of the long side from the bottom right vertex; see points I and J in Fig.10.17.

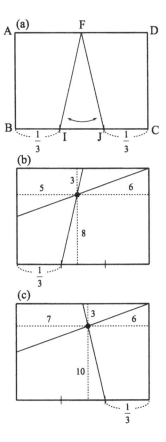

Fig. 10.17
(a) To obtain the 11-parts and 13-parts division points "pendulum" lines from midpoint F to the trisection points of the lower side are used

(b) The 11-parts division point (using the left trisection point) gives a ratio of 5 : 6 in the horizontal direction left to right and 8 : 3 in the vertical direction upwards

(c) The 13-parts division point (using the right trisection point) gives a ratio of 7 : 6 in the horizontal direction and 10 : 3 in the vertical direction

These points are 11-parts division and 13-parts division points respectively. We can verify our observations by using a coordinate system as in previous discussions. This is left to the reader.

10.7 Another Method for Finding 11-parts and 13-parts Division Points

In the previous section, if we try to design a folding procedure on the basis of the discussion, a problem is how to obtain the trisection points I and J. Figure 10.2 describes three methods for finding a 3-parts division point; then folds can be made to obtain the trisection points on the sides. But these procedures are tedious, so we look for some other procedure.

Using computer software or by sketching we can construct a rectangle with the desired divisions; then we look for other lines in lieu of FI and FJ that intersect ED at a node. We examine lines through points on the upper side other than F. By experimenting, one likely setup was found for 11-parts division and another for 13-parts division (see Figs.10.20 and 10.22). We shall describe these setups and find out if indeed an equal-part division point is obtained.

For the 11-parts division, instead of using the midpoint of the upper side we take the points whose distance from the left vertex is $\frac{1}{4}$ of the length of the side. And join it to the right bottom vertex. Call this point K; see Fig.10.19.

We study the intersection of this line with ED. By using a coordinate system, we can locate the position of the intersection and find out for sure if it is an equal-parts division point.

Line ED: $$y = \frac{x + \sqrt{2}}{2\sqrt{2}} \qquad (1)$$

Line KC: $$y = -\frac{4x}{3\sqrt{2}} + \frac{4}{3}. \qquad (2)$$

Setting (1) = (2),

$$x = \frac{5\sqrt{2}}{11} \quad \text{and} \quad y = \frac{8}{11}$$

(see Fig.10.18).

Indeed, this gives us an 11-parts division point.

On the basis of the ratios given in Fig.10.19, a folding procedure can be carried out. This is left to the reader.

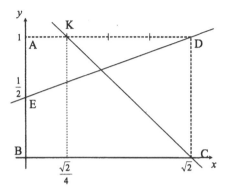

Fig. 10.18 Assuming the width is 1 and length is √2, the coordinates of the points are as shown in the figure.

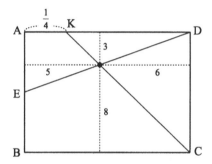

Fig. 10.19 The line joining K and the lower right vertex gives an 11-part division.

For the 13-parts division we use the point on the bottom side whose distance from the left vertex is $\frac{1}{8}$ of the side, call this point L, and connect it to the midpoint F of the top side. See Fig.10.21. The point cutting $\frac{1}{8}$ of the long side can be found by taking half of the $\frac{1}{4}$ – distance obtained by the method of Fig.10.10.

We calculate the position of the intersection of this line, FL with the line ED and find out for sure if it is an equal-parts division point.

We study the intersection of this line with ED. By using a coordinate system, we can locate the position of the intersection and find out for sure if it is an equal-parts division point.

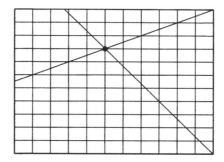

Fig. 10.20 The paper is folded into 11 equal parts horizontally and vertically.

The details of the computation are left to the reader. We arrive at

$$x = \frac{5\sqrt{2}}{13} \quad \text{and} \quad y = \frac{9}{13}.$$

Indeed, it is a 13-parts division point!

Look back at the 11-parts and 13-parts division points obtained in the previous section. Are they the same as those found here?

A natural follow-up study to the last situation would be to consider the pendulum line. Does this lead to an equal-parts division? The motivated reader may pause to find out.

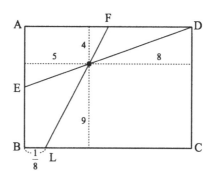

Fig. 10.21 The 13-parts division point is obtained by joining the midpoint F with the point L which cuts at $\frac{1}{8}$ of the length of the lower side.

INSPIRARATION OF RECTANGULAR PAPER 125

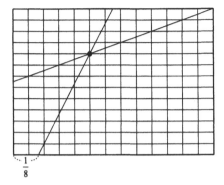

Fig. 10.22 Based on the point of intersection of Fig.10.21, the rectangle is divided into 13 equal parts horizontally and vertically.

10.8 Continue the Trend of Thought: 15-parts and 17-parts Division Points

Now we are coming closer to the Origami King's problem.

So far we have obtained methods for locating the 3-parts, 5-parts, 7-parts, 9-parts, 11-parts and 13-parts division points. Let us review them and look for patterns.

One observation concerns the slopes of lines which are pendulum pairs. In the 3-parts and 5-parts divisions the two lines through the midpoint of the upper side which are "pendulum pairs" (lines FB and FC in Fig.10.5) have slopes numerically equal but opposite sign.

In the 7-parts and 9-parts divisions the two lines through the midpoint of the upper side which are "pendulum pairs" (line FG in Fig.10.10 and line FH in Fig.10.13) have slopes numerically equal but opposite sign.

This relationship between slopes of one being the negative of the other is an algebraic way of describing the "pendulum" relationship. That is, the two lines are mirror images with respect to the vertical line through F.

Another observation is as follows. Note that the numbers are odd. So we hypothesize: the next two odd numbers are 15 and 17, might it be possible to find 15-parts and 17-parts division points by similar methods?

Still another observation concerns the lines through F. In all the procedures studied the line ED was constant. However, the line through the midpoint F varied; different reference points were located on the lower side BC. Observe:

For the 3-parts and 5-parts division use the vertices B or C.

For the 7-parts and 9-parts divisions use reference points whose distances from the respective vertices (B or C) is $\frac{1}{4}$ of the long side.

For the 11-parts and 13-parts division use reference points whose distances from the respective vertices is $\frac{1}{3}$ of the long side.

It appears that as the number of parts increase the reference point on the lower side becomes farther from the respective vertices: (i.e., 0, $\frac{1}{4}$, $\frac{1}{3}$). So we hypothesize: for the 15-parts and 17-parts division points might the distance of the reference points from the respective vertices be a fractional number greater than $\frac{1}{3}$?

To check our hypotheses we use computer software or sketch a figure. Let us begin with 15-parts division. To find a 15-parts division point we think of a fraction greater than $\frac{1}{3}$; try $\frac{3}{8}$.

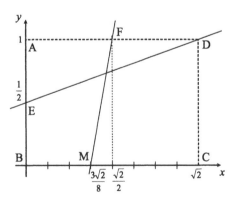

Fig. 10.23 The distance of point M from the origin B is $\frac{3}{8}$ of the length of BC. We need to determine if the intersection of ED and FM is a 15-parts division point.

Line ED: $$y = \frac{x + \sqrt{2}}{2\sqrt{2}} \qquad (1)$$

Line FM: $$y = \frac{8x - 3\sqrt{2}}{\sqrt{2}}. \qquad (2)$$

Setting (1) = (2),

$$x = \frac{7\sqrt{2}}{15} \quad \text{and} \quad y = \frac{11}{15}.$$

In other words, the point of intersection can be used to divide the rectangle into 15 equal parts, horizontally in the ratio 7 : 8 and vertically in the ratio 11 : 4 (Fig.10.25).

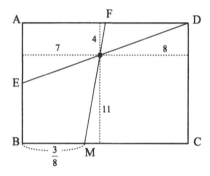

Fig. 10.24 The line joining F and M intersects the line ED at a 15-parts division point.

A folding procedure can now be carried out. The position of M may be found by first finding the rightmost quadrisection point, which cuts off $\frac{1}{4}$ of the side, then folding the remaining $\frac{3}{4}$ in half. The rest of the folding directions is left to the reader. Figure 10.25 shows the result of folding.

Next think of the pendulum pair; might it produce a 17-parts division? Let us find out.

If we "swing" line FM to the right, the new position of M will be at a point whose distance from C is $\frac{3}{8}$ of the long side. Call this point P; see Fig.10.26.

It can be confirmed that the intersection of lines FP and ED is indeed a 17-parts division point. Using a coordinate system we obtain

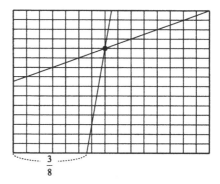

Fig. 10.25 Based on the intersection point of Fig.10.24, the rectangle is divided into 15 equal parts horizontally and vertically.

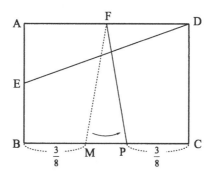

Fig. 10.26 The line FM and its pendulum pair, FP.

Line ED: $$y = \frac{x + \sqrt{2}}{2\sqrt{2}} \qquad (1)$$

Line FP: $$y = \frac{8x + 5\sqrt{2}}{\sqrt{2}}. \qquad (2)$$

Setting (1) = (2),

$$x = \frac{9\sqrt{2}}{17} \quad \text{and} \quad y = \frac{13}{17}$$

(see Fig.10.27).

Thus the point of intersection can be used to divide the rectangle horizontally in the ratio 9 : 8 and vertically in the ratio 13 : 4.

INSPIRARATION OF RECTANGULAR PAPER 129

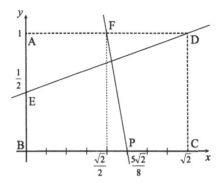

Fig. 10.27 Assuming the width is 1 and length is $\sqrt{2}$, the coordinates of the points are as shown in the figure.

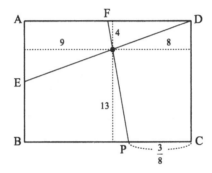

Fig. 10.28 The point of intersection of ED and FP can be used to divide the rectangle horizontally into 17 equal parts in the ratio 9 : 8 and vertically in the ratio 13 : 4.

Regarding the folding procedure, clearly it becomes increasingly difficult to fold as the number of folds increase. The strips tend to become skewed in spite of one's intention to fold along parallel lines. Therefore one must be very careful; when folding one crease, first smooth the crease lightly with the bulb of the finger, then check and correct deviations by aligning the left and right corners. And finally make a sharp crease using the fingernails.

If we make our creases carefully, we have solved the problem of the Origami King!

130　　　　　　*Dividing a Rectangular Sheet of Paper*

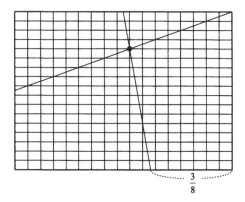

Fig. 10.29　Based on the intersection point of Fig.10.28, the rectangle is divided into 17 equal parts horizontally and vertically.

10.9　Some Ideas related to the Ratios for Equal-parts Division based on Similar Triangles

In most of the discussions so far, to obtain an equal-parts division point the line ED and the midpoint F were used; and variable lines were made through F intersecting the opposite side of the rectangle. From the diagrams of the grids or of computer software, we can understand that different equal-parts division points are possible.

With these thoughts in mind, Ms. Keiko Odagiri, a middle school teacher in a Japanese school, pointed out that another viewpoint might be to consider the similarity ratios of the sides of similar triangles.

Her observations progressed as follows. In Fig.10.30 we label the intersection of BC and the line through point F; call it G. Also we label the point of intersection of FG and ED; call it P.

(1)　　In the diagram extend the side BC along the x axis, and extend line ED. These lines will intersect; call the point of intersection H. We see that　△FDP ~ △GHP.
This ratio might be related to that in equal-parts vertical division In order that ratios are in simplest form, we shall confine divisions to an odd number of parts.

(2) For the fixed line ED, the similarity relationship holds regardless of the position of F. Therefore, it is possible to choose point F and G corresponding to $\frac{1}{2^k}$ of the long side or some other fraction easily obtained by folding.

(3) The idea described in (2) holds regardless of the position of the fixed line; in particular if the fixed line is a diagonal.

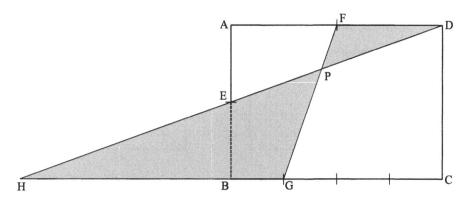

Fig. 10.30

The following sketches illustrate the above ideas. Be careful: it is important to take note of the directions of the ratios — i.e., left-to-right or right-to-left, and upwards or downwards.

The above sketches bring out another interesting relationship, described by a Mr. Yoichi Otani (as you can see, the author had a number of acquaintances with whom he discussed his origamics explorations). As shown in the sketches, Mr. Otani used the diagonal from lower left to upper right (the positive diagonal) as the fixed line and the lower right vertex as an endpoint of the second line. Then he set different positions of F and studied the corresponding ratio of heights. He observed:

if F cuts off $\frac{1}{2}$ of the side right-to-left, the two lines create a 2 : 1 division point;

132   *Dividing a Rectangular Sheet of Paper*

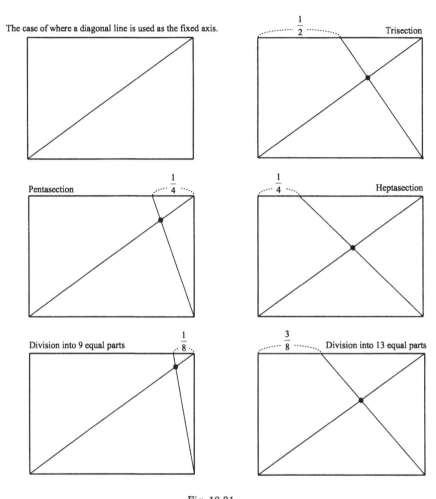

Fig. 10.31

if F cuts off $\frac{1}{3}$ of the side right-to-left, the two lines create a 3 : 1 division point;

if F cuts off $\frac{1}{4}$ of the side right-to-left, the two lines create a 4 : 1 division point;

⋮

if F cuts off $\frac{1}{k}$ of the side right-to-left, the two lines create a

$k : 1$ division point.

Furthermore,

if F cuts off $\frac{j}{k}$ of the side right-to-left, the two lines create a $k : j$ division point.

Another interesting relationship is embodied in Fig.10.32. Here the principal diagonal is also used as the fixed line, but the division point is on the lower side.

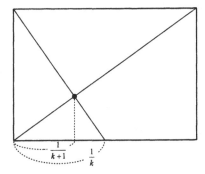

Fig. 10.32 An interesting mathematical principle can also be found in the intersection of a line from the upper left corner and diagonal lines.

As this figure shows, there is still undoubtedly great potential for further development of the topic of dividing paper into an odd number of equal parts.

And while the discussions in this section related to dividing in one direction only (i.e., vertical), it is possible that further studies would lead to two-direction division (i.e., vertical and horizontal) and therefore to the main idea of this chapter.

And therefore this section is open-ended.

10.10 Towards More Division Parts

So far we have been studying odd number of equal-parts divisions. If we wish to pursue this research further, we note that, except for 9 and 15, the numbers studied are prime. Furthermore, if prime number division is possible, it might also be possible to design other odd number divisions as combinations of prime number procedures. With these thoughts in mind the author decided to continue the study with prime numbers.

After 17 the next prime numbers are 19, 23, 29, 31, and so on. However, in actual folding using no tools, there are limits related to manual dexterity, visual acuity, and the properties of the paper. As of this writing 31 equal parts with A4 paper appears to be the limit of the author's ability. However, because calculation and computer display analysis have shown that it is possible to continue beyond this; and therefore the author hopes that readers with perseverance and skillful fingers will take up the challenge.

10.11 Generalizing to all Rectangles

Incidentally, equal-parts folding by the methods described here is not limited to paper whose sides are in the ratio such as the A4 and B5 sizes. The methods are effective for rectangles with sides in any ratio $1 : a$. The methods also apply when the ratio is $1 : 1$ - that is, a square.

Even if the Origami King insists on square paper, undoubtedly he will be convinced by the solution to his problem which we have discussed.

Truly, origamics has much potential for discovery, examination and confirmation of mathematical principles; and at times also the potential of emotions of surprise and wonder. I hope that the readers will make good use of origamics for learning.